Lipid migration in crystalline fat suspensions:
Investigation of possible pathways and mechanisms

Lipid migration in crystalline fat suspensions: Investigation of possible pathways and mechanisms

Vom Promotionsausschuss der

Technischen Universität Hamburg-Harburg

zur Erlangung des akademischen Grades

Doktor-Ingenieur (Dr.-Ing.)

genehmigte Dissertation

von

Svenja Kristin Lügger

aus

Hamburg

2017

1. Gutachter: Prof. Dr.-Ing. habil. Dr. h.c. Stefan Heinrich

2. Gutachter: Prof. Dr.-Ing. Irina Smirnova

3. Gutachter: Prof. Dr.-Ing. Stefan Palzer

Tag der mündlichen Prüfung: 16. März 2017

Bibliografische Information der Deutschen Nationalbibliothek
Die Deutsche Nationalbibliothek verzeichnet diese Publikation in der Deutschen
Nationalbibliografie; detaillierte bibliografische Daten sind im Internet über
http://dnb.d-nb.de abrufbar.
    1. Aufl. - Göttingen: Cuvillier, 2017
    Zugl.: (TU) Hamburg-Harburg, Univ., Diss., 2017

Cover-Foto: Heiner Müller-Elsner, DESY

© CUVILLIER VERLAG, Göttingen 2017
    Nonnenstieg 8, 37075 Göttingen
    Telefon: 0551-54724-0
    Telefax: 0551-54724-21
    www.cuvillier.de

    ISBN 978-3-7369-9517-8
    eISBN 978-3-7369-8517-9

Abstract

Lipid migration in crystalline fat suspensions: Investigation of possible pathways and mechanisms

by Svenja LÜGGER

The formation of visible white spots or a greyish haze on the surface of chocolate is a major quality issue for the confectionery industry. This phenomenon is the so-called chocolate fat blooming and results in consumer rejections and thus large sales losses. The chocolate industry is therefore interested in understanding the formation mechanism and its driving force to find solutions to prevent chocolate fat blooming. Thus, several researchers have investigated the phenomenon since many years. However, until now no study presents a complete explanation of the mechanism leading to its formation. One of the main theories for the formation of chocolate fat blooming is migration of lipids to the surface with subsequent recrystallization. Possible driving forces are diffusion of lipids and capillary pressure. Lipid transport driven by capillary pressure requires capillaries such as cracks or a pore network through which lipids migrate. Thus, micro- and nanostructure influence the migration pathway and mechanism. This thesis discusses potential pathways and investigates possible driving forces for lipid transport through chocolate. Synchrotron based microtomography revealed the inner chocolate microstructure. Moreover, structural changes due to lipid transport were tracked with small angle X-ray scattering. In addition, observations of macroscopic movement of stained lipid layers in different chocolate model systems gave further insight into the mechanisms. The experimental results indicate that lipid migration in chocolate involves different overlapping mechanisms. Small angle X-ray scattering manifested that lipids migrate through nanopores possibly driven by capillary pressure in the short-term. In the long-term lipids migrate through the matrix phase by dissolution of cocoa butter. Microtomography images of conventional chocolates confirmed the presence of voids, which might act as preferred pathway for lipid transport. Thereby, the production process and storage conditions significantly impact the porosity and thus migration mechanism and rate. The findings indicate that structure plays a major role and should be optimized by the production process to reduce lipid migration in chocolates. Further studies have to show if lipid migration is the reason for chocolate fat blooming.

Contents

List of Figures

List of Tables

Symbols

Symbol	Unit	Description	Equation
α	[-]	crystallized part	2.1
α_{max}	[-]	maximum crystallized part	2.1
Δc_A	[mol/m^3]	concentration gradient of component 'A'	2.4
Δk	[a.u.]	total scattering power less peak intensities	3.5
Δp	[Pa]	pressure difference	2.6, 2.8
Δx	[m]	distance	2.4
ϵ	n/a	fat property (e.g. mechanical, textural)	2.2
η	[Pa s]	viscosity	2.6, 2.10, 2.11, 3.8
Γ	[$\sqrt{m}$]	imbibition ability	2.11, 2.13
γ	[N/m]	surface tension of the liquid	2.8, 2.9,
λ	[m]	wavelength	2.3, 3.1
ν	[-]	Poisson's ratio	
ϕ	[-]	volume porosity (fraction)	2.12, 3.2, 3.8
ϕ_0	[-]	volume porosity at start of capillary rise	2.12, 2.13
ϕ_i	[-]	initial volume porosity	3.8
ψ	[°]	inclination angle of capillary	2.9
ρ	[kg/m^3]	density	2.9, 3.2
τ	[-]	tortuosity	2.12, 2.13
Θ	[°]	angle between incident X-ray beam and detector	3.1
θ	[°]	contact angle	2.9
θ_{CB}	[°]	contact angle of oil droplet on cocoa butter	3.6
θ_{glass}	[°]	contact angle of oil droplet on glass	3.6

Symbol	Unit	Description	Equation
A	[m^2]	contact area	2.14, 3.13
A_{cap}	[m^2]	cross section of capillary	3.11
$A_{cap,mol}$	[m^2]	cross section of capillary occupied with molecules	3.11
A_{mol}	[m^2]	cross section of molecule	3.11

a	$[\mathrm{m}^2/\mathrm{s}]$	constant a, which describes rate of migration	3.7, 3.9
a_0	$[\mathrm{m}^2/\mathrm{s}]$	initial value of constant a at start of simulation	3.14
a_{sim}	$[\mathrm{m}^2/\mathrm{s}]$	simulated constant a	3.13, 3.14
c	$[\mathrm{m}^2]$	constant c, which describes initial condition	3.7, 3.9
D	$[\text{-}]$	fractal dimension	2.3
D_A	$[\mathrm{m}^2/\mathrm{s}]$	diffusion coefficient of component A into B	2.5
D_{eff}	$[\mathrm{m}^2/\mathrm{s}]$	effective diffusion coefficient	2.4, 2.14,
D_{sim}	$[\mathrm{m}^2/\mathrm{s}]$	diffusion coefficient from simulated capillary rise	3.13
E	$[\mathrm{Pa}]$	Young's modulus	2.3
g	$[m/s^2]$	gravity	2.9
f_h	$[\text{-}]$	ratio of capillary radius and hydrodynamic radius	3.10
h	$[\mathrm{m}]$	migration height	3.7,
			3.8
h_0	$[\mathrm{m}]$	initial height	3.9
h_{max}	$[\mathrm{m}]$	maximum migration height	2.9
h_{exp}	$[\mathrm{m}]$	experimentally measured migration height	3.6
h_{sample}	$[\mathrm{m}]$	sample height	3.13
h_{sat}	$[\mathrm{m}]$	saturation height	3.13
h_{CB}	$[\mathrm{m}]$	adapted migration height based on θ_{CB}	3.6
I	$[\mathrm{a.u.}]$	scattering intensity	3.2 , 3.3
$I_{baseline}$	$[\mathrm{a.u.}]$	baseline of peak scattering intensity	3.4
I_{fit}	$[\mathrm{a.u.}]$	fitted scattering intensity	3.4
$\dot{J}_A$	$[\mathrm{mol}/(\mathrm{m}^2\ \mathrm{s})]$	diffusive flux	2.4
K	$[\mathrm{m}^2]$	hydraulic permeability	2.6, 2.12,
			3.8
k	$[\mathrm{a.u.}]$	scattering power	3.2, 3.5
k^*	$[\mathrm{a.u.}]$	total scattering power	3.3, 3.5
k_A	$[1/\mathrm{s}]$	overall rate constant of Avrami equation	2.1
k_{peak}	$[\mathrm{a.u.}]$	peak intensity	3.4, 3.5
L	$[\mathrm{m}]$	length	2.6
m_s	$[\mathrm{g}]$	mass at saturation	2.14
m_t	$[\mathrm{g}]$	mass at time t	2.14
n	$[\text{-}]$	dimensionless Avrami constant	2.1
n_{cap}	$[\text{-}]$	number of capillaries	3.11
p	$[\mathrm{Pa}]$	pressure	2.6, 2.8
			3.8
Q	$[\text{-}]$	diffusibility	2.5
q	$[1/\mathrm{m}]$	scattering vector	3.1, 3.2,
			3.3, 3.4

Symbol	Unit	Description	Eq.
q_{Flow}	$[\text{m/s}^2]$	flow	2.6
q_{min}	$[1/\text{m}]$	minimum scattering vector	3.3
q_{max}	$[1/\text{m}]$	maximum scattering vector	3.3
R_1, R_2	$[\text{m}]$	principal radii of curvature	2.8
r	$[\text{m}]$	radius of capillary	2.9, 2.12
r_0	$[\text{m}]$	initial capillary radius at start of simulation	2.13, 3.14
r_h	$[\text{m}]$	hydraulic/hydrodynamic radius of capillary	2.12, 2.13
r_l	$[\text{m}]$	radius on which liquid meniscus acts on	2.8, 2.13
r_{sim}	$[\text{m}]$	capillary radius based on simulation	3.14
SFC	$[\text{-}]$	solid fat content	2.2, 2.3
T	$[^\circ\text{C}]$	temperature	2.3
t	$[\text{s}]$	time	2.1, 2.14, 3.7
t_0	$[\text{s}]$	initial time	3.9
V	$[\text{m}^3]$	volume	2.14, 3.2, 3.13

1

Introduction

Chocolate blooming is one of the major problems in confectionery industry. It is manifested by the formation of white spots or a greyish haze on the chocolate surface due to large fat or sugar crystals. In case of fat crystals one refers to fat blooming and in case of sugar to sugar blooming. The crystals at the surface scatter the incident light and thus appear whitish (Altimiras et al., 2007, Lonchampt and Hartel, 2004, Rousseau and Smith, 2008, Rousseau and Sonwai, 2008). This makes the chocolate unappealing and leads to consumer complaints and subsequently large sales losses for the confectionery industry (Afoakwa et al., 2009a, Aguilera et al., 2004). Formation of sugar bloom is probably a result of recrystallization of dissolved sugar on the chocolate surface. Thereby, moisture dissolves sugar due to for example storage at low temperatures (e.g. fridge) and sugar recrystallizes at the surface. The formation of fat bloom might result from migration of lipids through the chocolate with subsequent recrystallization on the surface (Aguilera et al., 2004, Altimiras et al., 2007, Ghosh et al., 2002, Hartel, 1999). Many research studies aiming to understand blooming of chocolate have been published. Chapter 2 gives an overview of published studies related to chocolate fat blooming and lipid migration in chocolate. But the exact mechanisms are still not fully understood (Aguilera et al., 2004). However, a better understanding of pathways and driving force is a prerequisite to find ways to avoid lipid migration and thus the associated fat bloom formation. There are different possible pathways in chocolate which is composed of particles embedded in a crystalline fat matrix. Thus, migration might take place in the matrix phase, the network of particles or at the interface of particles and matrix. Two main mechanisms are widely discussed in literature: Diffusion as well as convective flow driven by capillary pressure are potential migration mechanisms (Hartel, 1999, Lonchampt and Hartel, 2004). The aim of this thesis is to identify preferred pathways and evaluate possible migration mechanisms. Thereby, the influence of micro- and nanostructure on lipid migration in chocolate is a main focus, because it determines both, the pathway and the mechanism. By investigation of possible crevices and pores in chocolate, convective flow as a possible mechanism for lipid migration might be excluded or identified as the main transport mechanism. The methods and materials are introduced in Chapter 3.

To identify preferred pathways small angle X-ray scattering tracks migration of oil into porous chocolate powder components in-situ. Small angle X-ray scattering delivers information on a molecular and nanoscale without sample destruction. Besides the evaluation of preferred pathways, structural changes induced by oil migration in chocolate model systems can be investigated. Thereby, synchrotron radiation enables a high spatial and time resolution and is thus a promising technique. Chapter 4 presents and discusses the results of the small angle X-ray study.

Apart from structure features on a molecular and nanoscale, the particle arrangement within chocolate and thus surface area is of high interest for explanation of potential transport pathways. Therefore, further information about the microstructure of chocolate might reveal possible pathways and validate the proposed mechanisms. X-ray tomography is a technique which images the interior structure of a material without sample destruction. The image capture is based on differences in electron density of the different components. The use of synchrotron radiation leads to improved density contrast and consequently enables distinction of particles and imperfections such as cracks, crevices and voids in conventional chocolates. The visualization of chocolate model systems (the fat matrix without particles, pure cocoa butter, and samples with varying particle amount suspended in cocoa butter) supports the understanding of microstructure of chocolate. Finite element method simulations of the solidification process of chocolate gives possible explanations for the origin of cracks and crevices found in chocolate. Chapter 5 is about the structure analysis of chocolate with X-ray tomography.

The impact of microstructure on migration rate is analyzed with macroscopic observation of migration in different chocolate model systems. Thereby, the influence of tempering, which is controlled crystallization, is of high interest. Tempering leads to a more homogeneous and denser structure. In contrast, no tempering is characterized by a higher porosity and possibly inclusions of liquid and thus mobile lipids. A combination of the data from tomography and small angle X-ray scattering can explain the different migration rates found in the macroscopic samples. Based on these findings potential pathways and migration mechanisms can be evaluated. Chapter 6 examines the results from the macroscopic migration experiments. The thesis concludes with Chapter 7, which summarizes the results and proposes a migration model.

2

Theoretical background

The following chapter gives the theoretical background to lipid migration in chocolate. Therefore, first, the material chocolate is introduced in general followed by a more detailed description of the main component cocoa butter. Thereby, the composition, crystallization and the resulting microstructure of cocoa butter are presented. Subsequently, chocolate blooming and mass transport mechanisms are introduced. Finally, published hypothesis of lipid migration in chocolate are described.

2.1   Chocolate

Chocolate, a well-known confectionery product, is characteristic for both its flavor and texture. Thus, chocolate production aims to develop these two attributes to ensure the characteristic chocolate taste experience. Thereby, one of the main properties is that chocolate appears as a solid piece at room temperature but melts during consumption (at body temperature which is about 37 °C). From a material science perspective, chocolate is a suspension with particles surrounded by a semi-liquid crystalline fat matrix, mainly cocoa butter (Figure 2.1). Particles are densely packed with a particle concentration of about 70 wt% (Rousseau and Smith, 2008). The particles are usually sugar, cocoa solids and, in case of milk chocolate, milk powder. To adjust rheological properties an emulsifier which is often lecithin is added to the chocolate recipe (Lonchampt and Hartel, 2004). A mercury porosimetry study revealed that the matrix of chocolate is porous filled with liquid cocoa butter fractions (Loisel et al., 1997). Thus, lipids might migrate to the chocolate surface through the continuous fat phase (Figure 2.1 a), at the interface of the particles and matrix phase (Figure 2.1 b) or through the network of particles (Figure 2.1 c).

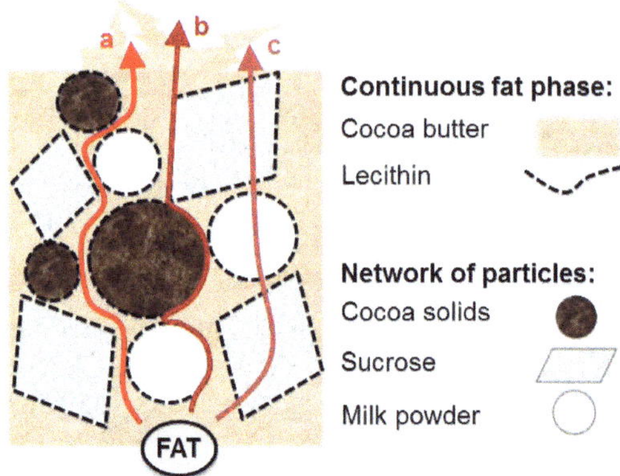

FIGURE 2.1: Chocolate structure: Sucrose, cocoa solids and milk powder particles surrounded by lecithin, an emulsifier, and embedded in a continuous cocoa butter matrix. Possible migration pathways of a lipid molecule within chocolate are a) through the fat matrix, b) at the interface of particles and fat matrix and c) through the network of particles.

A wide range of chocolate products exists on the market ranging from plain or filled chocolate bars over ice cream additives and pralines with fillings to bakery products. Chocolates can be subdivided into white, milk and dark chocolate. The latter typically contains 20 to 55 wt% of sugar, 45 to 80 wt% of cocoa mass and up to 5 wt% of additional cocoa butter as well as up to 0.5 wt% of lecithin. In contrast, milk chocolate usually consists of less cocoa but 12 to 18 wt% of non-fat milk solids and 3.5 to 6 wt% of milk fat. White chocolate typically lacks any cocoa mass and cocoa powder, but contains up to 50 wt% sugar, a third by weight milk powder and 22 wt% to 30 wt% cocoa butter and some milk fat resulting in a total fat content of 29 wt% to 40 wt%. (Wohlmuth, 2009)

Generally, the first step in chocolate production is to mix the ingredients, which are cocoa mass[1], sugar, fat and optionally milk powder. A size reduction step follows to decrease particle size to less than 40 to 20 µm so that single particles cannot be detected on the tongue and to get a smooth chocolate texture[2]. Typical apparatus for grinding are three- or five-roll refiners. The subsequent step is the so-called conching where the chocolate mass is constantly under agitation at elevated temperatures[3]. The aim of this step is to develop flavor and texture by removal of undesired acidic and astringent compounds plus to ensure a homogeneous coating of particle surfaces with fat (Beckett, 2009). These are the basic steps for production of chocolate mass, which appears as a viscous suspension. To form a final product with a good snap, gloss and stability, a

---

[1]Cocoa mass is produced from grinding cocoa beans which are fermented, dried, cleaned and roasted followed by separation of shells beforehand (Beckett, 2009)

[2]The exact size depends on taste and type of chocolate (Beckett, 2009).

[3]The actual temperature depends on the product (Beckett, 2009).

controlled solidification step is essential. For that purpose, the chocolate mass undergoes tempering, a precrystallization step, and subsequently solidifies under controlled cooling[4].

The surface of chocolate has a complex topography having an irregular texture with sizes of mostly less than 3 µm and many deep pores with a size of about 3 to 6 µm in diameter and various morphologies (Rousseau, 2006, Rousseau and Smith, 2008, Smith and Dahlman, 2005). The surface of fresh chocolate is relatively smooth and the pores are heterogeneously distributed over the surface. Dahlenborg et al. (2011, 2012) found in addition to pores also protrusion of various sizes from about a micron to about 20 µm which change over storage time into "ridge-like structures" on the surface of white chocolate pralines. The surface structures partly go at least 10 µm into the chocolate and some contained fat and others were empty spaces covered with fat (Dahlenborg et al., 2012). Loisel et al. (1997) confirmed the presence of pores in dark chocolate with mercury porosimetry. They found empty spaces of 1 to 4 %[5] which are partly filled with liquid fat. Thus, they propose that chocolate structure is a network of cavities with some liquid fat inside. Rousseau and Smith (2008) imaged the cross section of chocolate which has been cut prior to analysis with scanning electron microscopy. They visualized empty pores with a diameter of about 250 microns and channels going more than 100 microns through the entire chocolate sample.

2.2 Cocoa butter

The characteristic properties of chocolate, namely its texture and melting behavior, is strongly influenced by the fat matrix, which is mainly cocoa butter. The structure of cocoa butter influences the physical and crystal properties of cocoa butter such as melting behavior, crystalline form and strength (Aguilera, 2005, Campos et al., 2010, Rønholt et al., 2013, van Malssen et al., 1996). Physical, thermal and mechanical properties then again determine functional and sensory characteristics of the fat mixture and thus consumer experience (Campos et al., 2010, Sato, 1999). The cocoa butter's properties are strongly related to its structure which is determined by the arrangement of molecules. The molecular structure can be subdivided into the shape of the fat molecules (Figure 2.2, Level 2), the packing of the same (Figure 2.2, Level 3) and finally the crystal shapes (Figure 2.2, Level 4) forming crystal grains. The latter interact and finally form a crystal network, which gives cocoa butter the characteristic properties. Narine

---

[4]Further details on crystallization and tempering can be found in section 2.2.2, page 8 for crystallization in general and page 17 for the tempering process.

[5]The exact porosity was dependent on the solidification process of the chocolate mass and fat content. For well-tempered chocolate with 31.9 % cocoa butter they measured a porosity of 1 %, whereas it increased to 2 % at fat content of only 29.5 % and to 4 % for over-tempered samples.

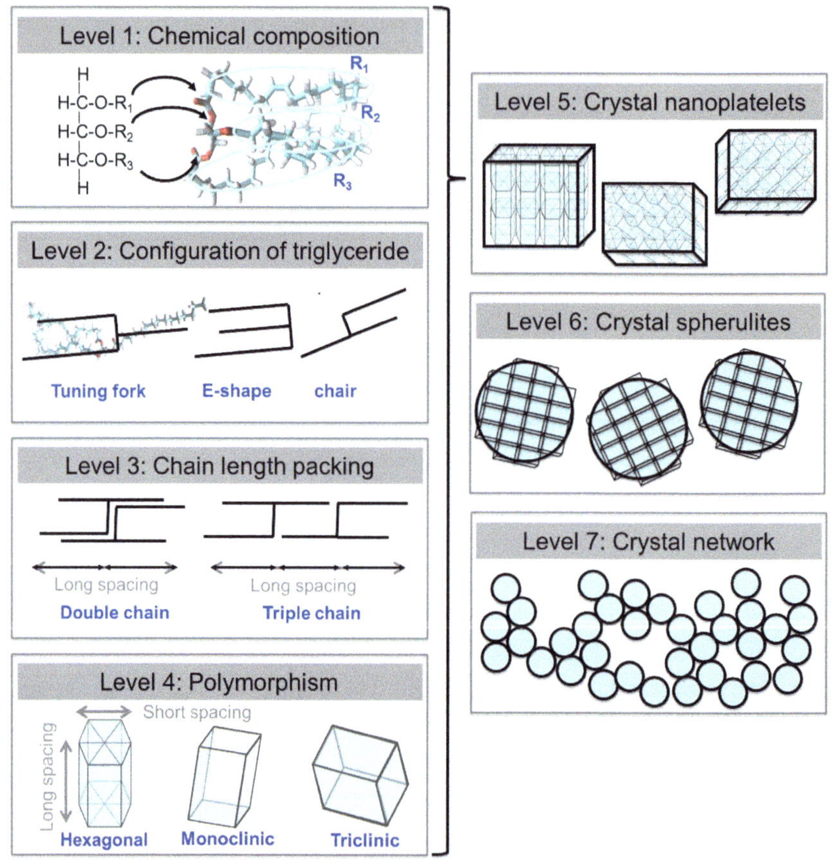

FIGURE 2.2: Different levels of structure of cocoa butter.

and Marangoni (1999b) describe the cocoa butter fat network structure as that of colloidal gels (Figure 2.2, Level 5 to 7) . The shape, packing and crystal structure depend on the composition (Figure 2.2, Level 1) alongside with the process conditions (mainly temperature and agitation during solidification) (e.g. Lutton, 1950, Rønholt et al., 2013, Sato, 1999). The different levels forming the cocoa butter structure are discussed in more detail in the following.

2.2.1 Chemical composition

Cocoa butter is a fat mixture comprised of mostly triglycerides (triacylglycerols), which are esters of glycerol and fatty acids (Figure 2.2, Level 1) with a few other minor components (Himawan et al., 2006, Lipp and Anklam, 1998). The chemical composition of cocoa butter is dependent on the origin, the harvest time and plant (see for example Figure 2.3). About 80 % of the cocoa butter are triglycerides with an unsaturated fatty acid in between two saturated fatty acids. More than 95 % of the fatty acids are oleic (O), stearic (S) or palmitic (P) (Beckett, 2000, Metin and Hartel, 2005). Oleic acid is composed of 18 carbon atoms with 1 unsaturated bond (C18:1 with an unsaturated bond at C9), stearic acid has 18 carbon atoms as well but no unsaturated bond (C18:0), same as palmitic acid which has 16 carbon atoms without any unsaturated bond (C16:0) (Campos et al., 2010). The saturated fatty acid chains are linear and have a kink at positions with double bonds (Himawan et al., 2006). Cocoa butter in chocolate contains palmitic, stearic and oleic acids in nearly equivalent amounts (Rothkopf and Danzl, 2015). The main triglycerides in cocoa butter are palmitic-oleic-palmitic (POP), stearic-oleic-stearic (StOSt) and palmitic-oleic-stearic (POSt) triglyceride. These three triglycerides make up about three quarter of components in cocoa butter (Figure 2.3) (Lovegren et al., 1976, Torbica et al., 2006, van Malssen et al., 1996). Their melting

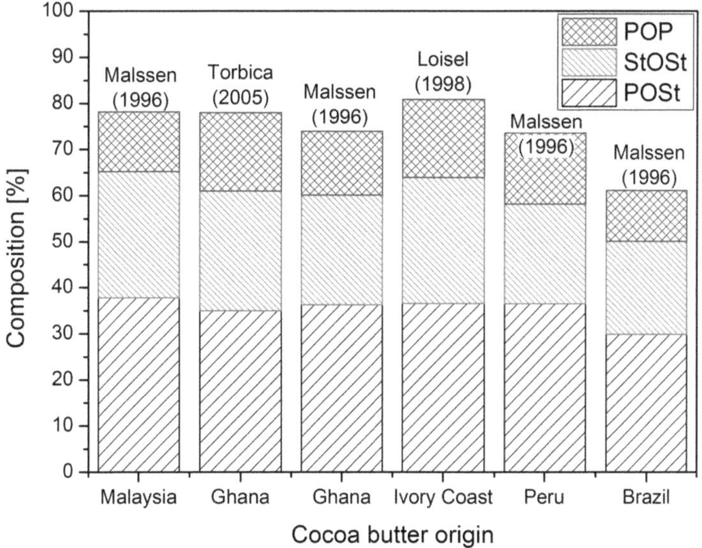

FIGURE 2.3: POP, POSt and StOSt amount in cocoa butter from different origins published in Torbica et al. (2006), Lovegren et al. (1976), van Malssen et al. (1996).

range is at about room temperature. Only about 1 % to 2 % of cocoa butter is all saturated fatty acids and thus melts at higher temperatures (Beckett, 2000). Besides the main triglycerides, 5 % to 20 % of cocoa butter fats have a saturated and two unsaturated oleic acids like stearic-dioleic (StOO) and palmitic-dioleic (POO), which are liquid at room temperature (Beckett, 2000). In addition, fats like cocoa butter contain di- and monoglycerides as well as phosphoric acids as other minor components (Beckett, 2000, Metin and Hartel, 2005).

2.2.2 Crystallization and structure

Cocoa butter solidifies in a crystalline form which gives the fat mixture the typical structure (e.g. Metin and Hartel, 2005, Sato, 2001). The crystallization process determines for example the mechanical (such as hardness) and visual (e.g. gloss, color, surface appearance) properties of cocoa butter induced by a variation in microstructure (surface as well as internal network structure) (Afoakwa et al., 2008a). Thus, the following aspects are shortly introduced: First, kinetics of cocoa butter crystallization are presented. Secondly, the packing of triglyceride molecules and formation of fat networks are discussed. Thirdly, the solid fat content, which is an important property of cocoa butter, is reviewed. The exact temperature profile of the solid fat content depends on a variety of conditions, such as chemical composition of cocoa butter, history and time effects and cooling conditions. And finally, the influence of minor components on crystallization is shortly presented in this section.

2.2.2.1 Kinetics of cocoa butter crystallization

Cocoa butter is liquid at temperatures above approximately 40 °C (Figure 2.5). At lower temperatures fats are in an ordered structure even in its liquid phase. This also leads to a memory effect according to Hernqvist (1990), Metin and Hartel (2005). So, fats might solidify in the same crystalline structure when re-cooled after melting at a lower temperature. Loisel et al. (1998) found with differential scanning calorimetry (DSC) and X-ray diffraction (XRD) a liquid-crystal organization with crystalline parts and less organized structure resulting in diffuse scattering. van Malssen et al. (1996) explained the memory effect with high stability of StOSt crystallites, being according to them the most important triglycerides for the memory effect, with melting temperatures of more than 40 °C (Sato et al., 1989).

The crystallization process starts with nucleation which is the formation of small crystals in the solution which subsequently grow (Metin and Hartel, 2005). Nuclei form due to accumulation and aggregation of molecules in the liquid phase (so called homogeneous nucleation). Nucleation can be facilitated by external nuclei such as foreign particles or walls (so called heterogeneous nucleation). Its rate depends on temperature profile and saturation but also on the polymorphic crystal form and determines number and size of crystals (Metin and Hartel, 2005).

Once nuclei are present the crystals grow by further aggregation of triglyceride molecules from the liquid phase. Molecules from the liquid phase diffuse and attach to the surface of the crystals to grow. According to the review by Metin and Hartel (2005) the attachment is more likely for similar molecules. Crystal growth ends when there is no more driving force for crystallization which it is at phase equilibrium of the system (Metin and Hartel, 2005). Adam-Berret et al. (2011) first observed melting of small crystals which subsequently recrystallize on larger ones due to Ostwald ripening. The higher the solid fat content, and thus the less liquid fat, the more pronounced was the effect of melting and recrystallization of small crystals.

A possible way to model crystallization kinetics is the Avrami equation (Avrami, 1939, 1940, 1941), which has been applied to cocoa butter crystallization by e.g. Le Reverend et al. (2009), Metin and Hartel (2005). Ziegleder (1990) analyzed crystallization with differential scanning calorimetry (DSC). According to him it is possible to linearize the DSC measurement curves with the Avrami equation:

$$1 - \frac{\alpha}{\alpha_{max}} = e^{-k_A \cdot t^n} \tag{2.1}$$

with the time t, α which is the crystallized part and α_{max} which is the maximum crystallized part. k_A is the rate constant and n a dimensionless constant. The Avrami equation describes the growth of crystalline spherolithes in polymers and gives physically relevant parameters such as a rate constant and spatial orientation. The influence of the origin, degree of ripeness, quality condition and temperature on the crystallization time of cocoa butter was investigated by Ziegleder (1990). The crystallization process is dependent on the origin, degree of ripeness and quality condition. Ziegleder (1990) used isothermal conditions in the DSC to get access to the fat crystallization. $lg(-ln(1 - \frac{\alpha}{\alpha_{max}}))$ can be plotted against $lg(t)$ which can be applied to the DSC data to determine the Avrami constants of Equation 2.1. Ziegleder (1990) found out that the crystallization time t showed a large range for cocoa butters with different origins. He explains the variations with the different compositions of fatty acids and triglycerides (Figure 2.3, Level 1). In general, diglyceride (such as present in cocoa with mold infestation) inhibit crystallization[6]. The tested cocoa butter from Malaysia was fastest compared to Nigeria and Bahia (which was slowest). Cocoa butter from southeast Asia is very hard (thus has a high melting temperature) because of many symmetrical and saturated triglycerides. This makes the development of a crystal easier and increases the rate of seed growth. There is a strong temperature dependence on the crystallization time. Whereby, the warmer the temperature, the slower the crystallization because of the exothermic nature of the

---

[6]Additional information on the influence of addition of other components such as minor components on crystallization are summarized in Section 2.2.2.6 on page 18.

crystallization process at temperature higher than 20 °C. At lower temperatures, seeding rate and growth are slowed down due to increasing viscosity. The seeding rate in the fat increases exponentially with linearly decreasing temperature (at temperatures larger than 20 °C). Furthermore, Ziegleder (1990) observed that there is a very small exothermic peak at the beginning of the isothermal phase probably due to the ingredients but this is not fully understood yet. Ziegleder (1990) describe the crystallization based on the Avrami equation as being a spherical process because the Avrami constant was $n = 4.4$ and a value of $n = 4$ is typical for three dimensional growth[7]. The deviation can be explained with non-homogeneous crystallization (Ziegleder, 1990).

The 3-dimensional network forms due to nucleation of triglycerides and subsequent growth as discussed in the part about structure of crystalline network (Section 2.2.2.3 on page 14). Thereby, particles inside the fat matrix influence the spatial distribution of fat crystal nuclei and crystallization pathway according to Rousseau and Sonwai (2008). They further explain that particles might act as sites for heterogeneous nucleation and possibly enhance growth by better heat and mass transfer[8].

Further information about solidification, phase transformation behavior and thermodynamic as well as kinetic aspects of fat crystallization can be found for example in Himawan et al. (2006), Metin and Hartel (2005), Sato (1999, 2001).

2.2.2.2 Packing of molecules in crystalline structure

The triglyceride molecules can pack into different geometrical crystal arrangements, denoted as polymorphic forms, which impacts fat properties such as melting profile (Himawan et al., 2006). Thus, from a crystallographic point of view, cocoa butter presents six known polymorphic forms, which are denoted either using Roman numerals (I-VI) or using the crystallographic nomenclature (γ, α, β_2', β_1', β_2, β_1) according to Lonchampt and Hartel (2004), Rousseau and Smith (2008), Wille and Lutton (1966). Both nomenclatures are evenly used in the literature and are interchangeable. In this thesis, the Roman numerals will be used to denote the cocoa butter polymorphic forms. van Malssen et al. (1996) instead claimed that only four distinct polymorphic forms exists with the other two being sub phases. They support their hypothesis with similar scattering patterns for the sub-phases and explain the differences in melting temperatures with a strong influence of composition. According to them, the different polymorphic forms are aggregations of individual crystallites with a distinct melting point each, which

---

[7]Further information on the application of the Avrami equation onto lipid systems can be found in Narine et al. (2006).

[8]The section about the influence of addition of other components such as minor components on crystallization is presented in Section 2.2.2.6 on page 18 and gives more examples on how other components such as minor ones or addition of particles influence the crystallization process.

individually depends on the triglycerides composition[9]. Molecular composition, temperature, pressure, rate of crystallization, impurities and shear rate influence the crystal configuration (Sato, 1999).

The fatty acid chains of the triglyceride molecules mainly orient into either a chair like or tuning fork shape (configuration) (Acevedo et al., 2011) but an E-shape is possible as well (Figure 2.2, Level 2). The triglyceride molecules can then pack into double or triple packings (Figure 2.2, Level 3) (Metin and Hartel, 2005). The chain packing can be distinguished with X-ray scattering according to their long spacings with larger long spacings for triple packing (e.g. Lutton, 1950)). Rønholt et al. (2013) state that the amount of unsaturated and saturated triglycerides influences the molecular symmetry. Saturated fatty acids are more symmetrical than unsaturated fatty acids. Thus, triglycerides with saturated fatty acids and high symmetry pack preferential in double length chains. And triglycerides which are not symmetric due to 2 or 3 different fatty acids or the ones with unsaturated fatty acids, which have a kink, mainly pack into triple packings (Lutton, 1950).

The arrangement into different crystallographic unit cells (Figure 2.2, Level 4) can be detected with scattering techniques and distinguished with their different melting temperatures with e.g. differential scanning calorimetry (DSC) (e.g. Cebula and Ziegleder, 1993, Wille and Lutton, 1966, Ziegleder, 1990). Scattering reveals the geometric orientation of the molecules within the crystal lattice. The unit cell of triglycerides can be characterized by its short and long spacing whereby both define the unit cell's short and longer dimensions (Figure 2.2, Level 3 and 4). The long spacing provides information on packing length (triple or double chain configuration, Figure 2.2, Level 3).

Cocoa butter molecules can arrange into 3 different geometric unit cells, denoted as alpha α, beta β and beta prime β'. Unit cells with equal geometric arrangement but differences in melting temperature are denoted with subscripts 1 and 2 whereby 1 is the one with a higher stability than 2. According to Wille and Lutton (1966), beta is a form with a short spacing of 4.6 Å and highest melting temperature, sub-beta melts just some degrees before the beta form and has a short spacing slightly larger than beta at about 4.7 Å. The short spacings of beta prime (an intermediate melting form) are at 4.2 Å and 3.8 Å. The alpha polymorphic crystalline form has a strong short spacing at 4.2 Å and sub alpha at values close to beta prime (Wille and Lutton, 1966). The alpha form has a hexagonal and the beta prime an orthorhombic unit cell (Lonchampt and Hartel, 2004). The beta form was usually associated with a triclinic structure (Lonchampt and Hartel, 2004) but more recent studies by van Mechelen et al. (2006a,b) propose that the beta polymorph of cocoa butter has actually a monoclinic crystal unit cell. Typical long

---

[9]More on the structure of the crystalline network can be found in subsection 2.2.2.3 on page 14.

spacings of form II to IV are between 45 Å and 49 Å (double chain, Figure 2.2, Level 3) and for forms V and VI 63 Å to 64 Å (triple chain) (Wille and Lutton, 1966).

Wille and Lutton (1966) and Loisel et al. (1998) among others identified the pure polymorphic forms of cocoa butter form II (alpha type), IV (beta prime type), V and VI (both beta type), whereby V and VI are closely related crystalline structures. van Mechelen et al. (2006b) solved the crystal structures of form V and VI with high resolution powder diffraction. They conclude that form V and VI are identical in the three packing structure and only differ in the symmetry between the three packings. Thus, they suggest that transformation from V into VI is a simple phase transition. Wille and Lutton (1966) did not clearly identify form I and III, which according to their study might be phase mixtures. They distinguished the forms by their short spacings which were found with scattering techniques. Loisel et al. (1998) propose a liquid-crystal structure for form I with sharp crystal peaks and diffuse scattering parts co-existing. The polymorph III was only observed shortly (Loisel et al., 1998).

Crystal formation depends on the temperature profile and agitation rate as well as composition of the lipid phase. The crystalline stability of cocoa butter increases from form I to VI, and form V is the main crystalline form in chocolate (Wille and Lutton, 1966). This polymorph presents the final desirable thermal and mechanical properties and gives rise to optimal chocolate appearance, and thus, it is preferred in consumer products. The formation of fat bloom[10] has been associated with a crystalline transformation of the cocoa butter from form V to VI, degrading the appearance of chocolate products. In addition, it has been reported that several cocoa butter polymorphs can coexist (Loisel et al., 1998). The different polymorphs distinguish themselves through varying stability, packing density and melting temperature. Besides the polymorphic form also the chemical composition determines the melting temperature, which depends on the cocoa butter origin. Figure 2.4 sums up the measured melting temperature ranges for cocoa butter and Table 2.1 gives a more detailed overview over experimental data from literature. The less stable polymorphic forms have a lower energy state and thus form faster but transform into more stable configurations with time (Lonchampt and Hartel, 2004, Marangoni and McGauley, 2003, Metin and Hartel, 2005, Sato, 1993). The polymorphic forms deviate in Gibb's free energy because of different molecular packings within the crystal lattice (Metin and Hartel, 2005). Therefore, fast cooling results in less stable polymorphic forms because the molecules lack time to order into stable, densely packed crystals (Metin and Hartel, 2005). But the unstable forms rearrange into more stable forms over time to get into a more favorable energetic state. The rearrangement is in general accelerated at higher temperatures of about 20 °C compared to slow transformation at less than 0 °C (Marangoni and McGauley, 2003). Pore et al. (2009) observed

---

[10]Further information on fat bloom can be found in Section 2.3 on page 21.

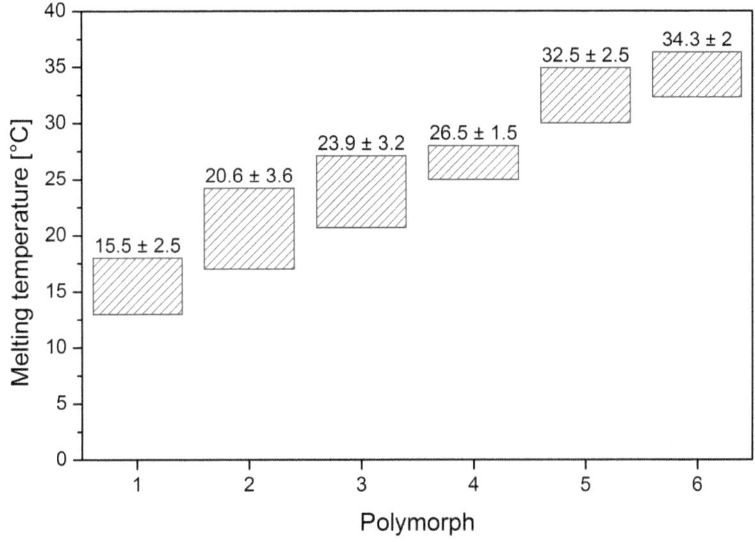

FIGURE 2.4: Range of melting temperature measured by Chapman et al. (1971), Duck (1964), Huyghebaert and Hendrickx (1971), Loisel et al. (1998), Lovegren et al. (1976), Merken and Vaeck (1980), Wille and Lutton (1966).

TABLE 2.1: Melting temperature of different polymorphic forms of cocoa butter reported by Chapman et al. (1971), Duck (1964), Loisel et al. (1998), Lovegren et al. (1976), Merken and Vaeck (1980), Wille and Lutton (1966) and Huyghebaert and Hendrickx (1971). Huyghebaert and Hendrickx (1971) indicated the ranges from maximum DSC melting to the temperature of the end of the melting peak with n.d.: not determined due to presence of crowded maximum.

		Melting	temperature	[°C]		
Polymorph	Duck (1964)	Wille, Lutton (1966)	Chapman et al. (1971)	Huyghebaert, Hendrickx (1971)	Lovegren et al. (1976)	Merken, Vaeck 1980
I	18.0	17.3		14.9-16.1	13	16-18
II	23.6	23.3		17-23.2	20	20.7-24.2
III		25.5	20.7	22.8-27.1	23	
IV	28.1	27.5	25.6	25.1-27.4	25	26-28
V	33.1	33.87	30.8	31.3-33.2	30	33.7-34.9
VI	34.4	36.3	32.3	n.d.-34.9	33.5	

polymorphic transition from form I over form II and III at 0 °C and further transformation to form IV and V at increased temperature of 24 °C of a small frozen cocoa butter droplet with X-ray diffraction within 2 hours. Phase transitions are similar to bulk cocoa butter but faster in the experimental set-up. They give a high density of nuclei and even temperature distribution as a reason. Marangoni and McGauley (2003), van Malssen et al. (1996) observed that the beta polymorph only formed from transformation of the beta prime form when cocoa butter is statically crystallized from the melt or as a solid state transition.

2.2.2.3   Structure of crystalline network

Besides the chemical composition, the microstructure of the fat mixture influence the properties of the product (Liang et al., 2006). Maleky and Marangoni (2011b) found that shear influences cocoa butter crystallization and consequently thermal and mechanical properties. The solid fat content and polymorphic forms of their samples were similar, thus they cannot explain the differences in material property. But, the arrangement of crystal components has an effect on strength of the material and melting temperature profile. They named crystal size, intercrystalline forces and spatial distribution of solids within the network as possible factors. Thereby, they observed that shear leads to a decrease in nanocrystal size. Liang et al. (2008) verified that microstructure of lipid systems significantly depend on formulation and processing conditions. Furthermore, they also found a significant influence of microstructure on compression modulus and thus mechanical properties.

Acevedo et al. (2011), Adam-Berret et al. (2011), Narine and Marangoni (1999b) point out that the structure of solid fat crystal networks, including cocoa butter, has different scale levels. The microstructure of cocoa butter fat network resembles the one of colloidal gels (Narine and Marangoni, 1999b). Thereby, the smallest units of the fat crystal network is according to Acevedo et al. (2011) nanoplatelets (Figure 2.2, Level 5). These nanoplatelets form larger fractal structures (Figure 2.2, Level 6) via van der Waals forces with sizes of about 6 µm (Narine and Marangoni, 1999c) which then arrange to a 3-dimensional network (Figure 2.2, Level 7) via aggregation with sizes of about 100 µm (Narine and Marangoni, 1999c). The microstructure is formed as an amorphous solid, whereas the primary particles at the intramicrostructural level are of fractal nature according to Narine and Marangoni (1999b). Adam-Berret et al. (2011) propose a minimum of three levels going from polymorphic arrangement of fat molecules on an angstrom level over crystal size in the micrometer range up to flocks found on a millimeter scale. These structures evolve over time and vary with different methods of crystallization (Adam-Berret et al., 2011). The interactions between the crystals determine the properties of the fat crystal network according to Rønholt et al. (2013). Thereby, process conditions such as cooling rate, shear and temperature during processing and chemical composition of the fat mixtures influence crystal size, shape and polymorphic structure, microstructure and consequently mechanical properties (Campos et al., 2010, Liang et al., 2006, Rønholt et al., 2013). Afoakwa et al. (2009a) found that hardness of under-tempered dark chocolate increased up to 5-fold during the first 3 to 4 days of storage, when a plateau is reached and no further change in hardness was observed. They explain the textural changes with restructuring and recrystallization of unstable fat crystals. The polymorphic form of the untempered cocoa butter transformed from mainly form IV, the main form directly after solidification, into V after a day of storage at ambient conditions and finally after 4 days into form VI. Liang et al. (2008) formulate a relationship of mechanical and textural properties ϵ (such as the compression modulus) with solid fat content SFC as well as microstructure in its simplest form being

$$\epsilon = f(SFC, microstructure) \qquad (2.2)$$

Thus, to get a quantifiable relationship, solid fat content and microstructure need to be expressed numerical. Solid fat content can be measured as a percentage and thus numerical expression is ensured[11]. Microstructure can be quantified among others by microstructure density, Euler characteristic, zero-crossing length, nearest-neighbor distance, fractal dimension (Liang et al., 2008). Liang et al. (2006) compared different approaches to quantify the microstructure of lipid systems. They imaged the samples with confocal scanning light microscopy and found microstructural units forming a network. Image analysis can be used to identify centroids of crystal flocs, microstructural units and their interfaces. This information can be converted into quantitative data such as microstructural units per volume, nearest neighbor features and fractal dimension. Liang et al. (2006) showed that the quantitative data can describe similarities as well as differences between lipid microstructures. Narine and Marangoni (1999b) propose a power law relationship of Equation 2.2 with the elastic modulus E as mechanical property.

$$E = \lambda \cdot SFC(T)^{1/(3-D)} \qquad (2.3)$$

The fractal dimension D describes the fractal arrangement of microstructural elements within the network and the λ parameter is a constant, which is according to Narine and Marangoni (1999c) a function of the size of the primary particles and on the interaction between them and is thus dependent on the lipid composition and crystal nature within the network. The parameter can be calculated with the Hamacker constant used to describe van der Waals forces and the diameter of the primary particles and their strain

---

[11]Section 2.2.2.4 on page 16 discusses the solid fat content in more depth.

as well as the distance between flocs (Brunello et al., 2003). The model is valid for fat networks with high solid fat contents of $SFC = 60$ to 100 wt%. This is the region where the links between the microstructures (Level 7, Figure 2.2) are stressed and not the ones within the microstructural elements (Level 6, Figure 2.2). This corresponds to the weak-link theory compared to the strong-link regime which is expected for lower fat contents Narine and Marangoni (1999b,c). Narine and Marangoni (1999b,c) support their model with experimental data from rheological measurements, hardness and microscopy.

2.2.2.4   Solid fat content

The solid fat content (abbreviated as SFC) is the mass fraction of solid in a solid-liquid fat mixture. It is most importantly determined by the chemical composition of the fat and temperature (Himawan et al., 2006). Cocoa butter is a fat mixture, which solidifies and melts, respectively, in a temperature range rather than a defined melting temperature. Therefore, cocoa butter is partly solid and partly liquid at room temperature with decreasing solid fat content with increasing temperature (Figure 2.5). Rønholt et al. (2013) describe that an increase in degree of unsaturation of the fatty acids and a decrease in chain length both result in a lower melting point. Wille and

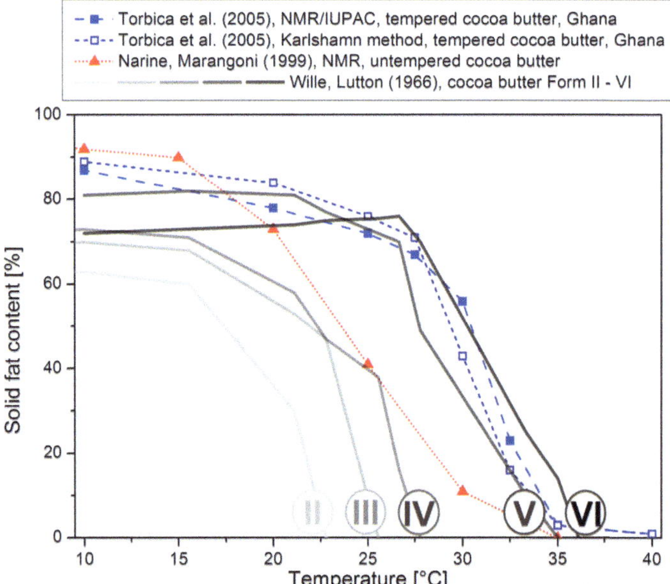

FIGURE 2.5: Solid fat content of different cocoa butter samples in dependence of temperature from literature Narine and Marangoni (1999a), Torbica et al. (2006), Timms (2002), Metin and Hartel (2005), Wille and Lutton (1966).

Lutton (1966) measured the SFC of cocoa butter in form II to form VI (Figure 2.5). In general, the solid fat content is higher for higher polymorphic forms because of increasing stability. The SFC of tempered cocoa butter, which has been solidified in a controlled manner[12] (Figure 2.5 by Torbica et al. (2006), Metin and Hartel (2005) and Timms (2002)) is in the range of the values for polymorphic forms V and VI by Wille and Lutton (1966). In contrast, the SFC of untempered cocoa butter as measured by Narine and Marangoni (1999a) is shifted towards the SFC of polymorph IV (Figure 2.5). As an additional effect, solid fat content may change over storage time. Rønholt et al. (2013) explain that post-crystallization during storage leads to an increase in solid fat content because of stronger interactions between crystals, which are mechanically interlinked, within the fat crystal network[13].

A common technique to measure solid fat content is nuclear magnetic resonance (NMR). Besides, cocoa butter origin and production process which determine the crystal form, different measurement protocols deliver different SFC data. Torbica et al. (2006) measured tempered cocoa butter with two measurement protocols, the IUPAC 2.150 and a newly developed protocol (Karlshamns). Both methods use nuclear magnetic resonance technique. They deviate in temperature profile during measurement with lower temperature and prolonged final stage for the latter one. The measured values deviate up to 13 %, whereas for lower temperatures of less than 30 °C the Karlshamns methods shows higher SFC and for higher temperatures it is opposite. Tabouret (1987) used electron spin resonance for determination of SFC to improve sensitivity. However, a tracer component needs to be added for electron spin resonance measurements.

2.2.2.5 Tempering process

The crystal structure of cocoa butter depends on the solidification procedure. The process of crystallization of fats in a controlled manner under specific temperature and agitation conditions is called tempering. The aim of the tempering process is to ensure a large number of very small crystals in the desired polymorphic form, which is form V in case of cocoa butter (Metin and Hartel, 2005). The process is essential to get the desired chocolate properties such as good snap, gloss and stability. Therefore, tempering targets at providing the right amount of crystal seeds in the correct polymorphic form from which crystals grow and solidification of the cocoa butter proceeds. Thus, seeds can be produced ex-situ and added to the chocolate mass (Zeng and Windhab, 1999). Another possibility is to first form seeds from the molted mass by targeted temperature profile under agitation. Therefore, the completely molten chocolate mass is cooled to the

---

[12]Further information on the tempering process can be found in Section 2.2.2.5 on page 17

[13]The forces between fat crystals constitute of irreversible primary bonds and mechanical interlinkage (Tempel, 1958) as well as reversible secondary van der Waals bonds (Haighton, 1965, Rønholt et al., 2013).

point of crystallization (typically about 27 to 30 °C)[14] which leads to formation of both stable and unstable crystals and is subsequently reheated above melting temperature of the unstable polymorphic forms but below melting temperature of the stable crystal forms (e.g. 30 to 32 °C) (Talbot, 2009a). Thus, seed crystals in a stable polymorphic form are present. Agitation is applied to ensure homogeneous distribution of seeds within the molten mass. Crystal growth proceeds from the present seeds leading to solidification of the cocoa butter mass.

The temper process influences the texture and appearance of the product. Insufficient (called under-tempering) but also excessive tempering (denoted as over-tempering) results in visual as well as mechanical quality defects (Afoakwa et al., 2008a). As an example, the microstructure of under-tempered dark chocolate shows voids within a coarse and heterogeneous structure (Afoakwa et al., 2009b). In contrast, well tempered samples had a homogeneous and dense crystal structure with inter-crystal connections (Afoakwa et al., 2009b).

2.2.2.6   Influence of addition of other components

Additional components in the triglycerides mixtures of cocoa butter can influence its crystallization and hence properties. For example, other vegetable fats or minor components such as mono- or diglycerides might both be incorporated into cocoa butter. Vegetable fats are sometimes added as an alternative of cocoa butter. They typically belong to either cocoa butter equivalents, substitutes or replacers with decreasing similarities to cocoa butter in physical and chemical properties (Talbot, 2009b). Smith et al. (2011) discuss the effects of additional components to fats on nucleation, crystal growth, morphology, heat capacity and polymorphic stability on a micron scale as well as visual aspects, melting behavior, post-crystallization, hardening and rheology, which are macroscopic properties. As a result they conclude that the impact of minor components on fat properties greatly depends on the similarity between the minor component and the fat with increasing impact for high similarity due to integration of the minor components into the fat crystal. Thus, one component can have varying effects on different fats. In addition, the cooling condition during solidification and concentration both influence the effect of additional components on fat properties (Smith et al., 2011).

The addition of other lipids might alter the crystallization process of cocoa butter and hence influences the solid fat content[15]. Campos et al. (2010) observed that addition of specific triglycerides alter the melting profile and thus also solid fat content of cocoa

---

[14]The exact temperature profile depends on the type and composition of the chocolate and of the cocoa butter.

[15]The basics of solid fat content are presented in Section 2.2.2.4 on page 16.

butter. They measured an increase in onset melting temperature for cocoa butter of about 5 °C because of addition of 1 % and 5 % tristearin (StStSt), a highly saturated and thus solid triglyceride. In contrast, addition of 1 % and 5 % trilinolein (LLL), a low melting component, only leads to a minor decrease in onset melting temperature of less than 1 °C. They explain this effect with alteration of crystallization and resulting crystal network due to addition of these two components. Addition of StStSt leads to fast undercooling of cocoa butter resulting in nucleation and crystal growth. However, StStSt is a relatively big molecule and thus produces higher disorganization inside the crystal lattice. This is the reason for slower polymorphic transformation from metastable polymorphic form to more stable forms. On the contrary, LLL only slightly effects crystallization but is incompatible with cocoa butter fats and thus does not co-crystallize. LLL increases the liquid fraction within the crystal network. This results in higher mobility and thus better rearrangement into a more stable crystal structure (Campos et al., 2010). Addition of fat mixtures with other chemical composition than cocoa butter also influence the crystallization behavior. Ribeiro et al. (2013) found that addition of 3 to 5 wt% triglycerides with high amounts of saturated fatty acids (named hard fats) to cocoa butter leads to higher melting temperatures and broadening of the melting range. Rothkopf and Danzl (2015) found an increase in crystallization speed, crystal growth rates and ability to crystallize of dark chocolate due to addition of hazelnut oil up to a concentration of 50 wt%. In contrast, butter fat slowed down crystallization and crystal growth rates. Furthermore, hazelnut oil promoted transformation from unstable to higher polymorphic forms whereas butter fat slowed down or even inhibits polymorphic rearrangement. However, the final solid fat contents of cocoa butter was independent of the investigated type of filling lipids. Rothkopf and Danzl (2015) found that the remaining cocoa butter can fully crystallize even with addition of up to 50 wt% of filling lipids when enough time for crystallization is ensured. Only the crystallization process differed for different types of added lipids. The underlying mechanisms might be that triglycerides from filling lipids are incorporated into the cocoa butter crystal lattice but are probably replaced during post-crystallization.

The concentration of additional components is not necessarily linearly related to the properties of the fat mixtures such as solid fat content but might follow an unpredictable trend (Lonchampt and Hartel, 2004). As an example, there might be no simple linear relationship between concentration of additional vegetable lipids and melting temperatures of the fat mixture due to formation of an eutectic systems. Thus, the mixture has a minimum in melting temperature which can be explained with two simultaneous effects (Rothkopf and Danzl, 2015). On the one hand dissolution of cocoa butter results in a decrease of solid fat content. However, pure dissolution would lead to a straight line between the solid fat content of the two components (cocoa butter and filling fat) without a minimum. Thus, a second effect has to take place which explains the minimum. A possible reason is the formation of an eutectic which leads to a further decrease up to a critical amount of additional fat (Rothkopf and Danzl, 2015). Kadivar et al. (2016) measured a significant decrease in solid fat content for cocoa butter blends with more than 25 wt% of sunflower oil accompanied with a strong softening effect. The melting behavior and solid fat content of blends up to 5 wt% sunflower oil was not significantly different to those of pure cocoa butter. However, addition of 5 wt% sunflower oil lead to a deceleration of post crystallization process during storage, which is observable by an increase in onset melting temperature after 4 weeks of storage for pure cocoa butter, but not for blends with 5 wt% sunflower oil.

In addition to the fat phase also particles might influence crystallization of the surrounding fat phase. Svanberg et al. (2011a,b,c, 2013) report that solid particles in the melt lead to alteration of the crystallization process[16], which is less pronounced in samples which are tempered with seeds. They imaged the crystal formation of cocoa butter from the melt to semi-solid during cooling with confocal laser scanning microscopy. Svanberg et al. (2011a) observed that the amount of stable crystals is higher in case of addition of cocoa compared to sugar particles to the cocoa butter matrix. The reason might be that more time for transformation of less stable polymorphic forms into more stable forms is available due to more rapid crystallization. Cocoa is more hydrophobic than sugar which is a possible explanation for the differences. This fact is strengthened by Svanberg et al. (2011a) experiments with addition of lecithin. Lecithin is a surfactant placed at the interface of particles and cocoa butter fat phase and by that might have facilitated heterogeneous nucleation. They found that addition of lecithin affected crystallization of cocoa butter in samples with sucrose and cocoa particles as well as in pure fat samples. In accordance to that, Afoakwa et al. (2008b) found that crystallinity as well as melting properties of cocoa butter suspensions depend on the fat and lecithin content. In general, they found that the higher the fat content (and thus less suspended particles), the higher the crystallinity and crystal size distribution. Furthermore, they found that the particle size of the suspended particles did not influence the crystallinity but the melting and mechanical properties (Afoakwa et al., 2008b, 2009a,b). They measured a decrease in temperature at the end of the melting peak with increasing particle size. But the onset and peak temperature as well as melting enthalpy were not influenced by the particle size distribution. Rousseau and Sonwai (2008) explain that crystallization is influenced by particles because of different spatial distribution of fat crystal nuclei and

---

[16]They used the number of nucleation sites and growth of crystal network as the two parameters to describe differences in the solidification of cocoa butter.

crystallization pathway. Thereby, the particles might act as nucleation sites. Furthermore, the addition of particles might facilitate the heat and mass transfer during the mixing and agitation process which might increase crystal growth. They propose that the surface of samples with particles is smoother because of promoted nucleation and faster growth resulting in smaller crystals. Rousseau and Sonwai (2008) suggest that the heterogeneously distributed particulate network, which is a result of the refining and tempering protocol, enables crystallization pathways of the fat phase and thus has an impact on morphology of chocolate.

2.3   Chocolate blooming

The formation of visible white spots or a grayish haze on the chocolate surface is called chocolate blooming. It is caused by either sugar (denoted as sugar blooming) or fat (named fat blooming) crystals on the surface which are in the micron size range and scatter the light (Altimiras et al., 2007, Lonchampt and Hartel, 2004, Rousseau and Smith, 2008, Rousseau and Sonwai, 2008). This leads to consumer rejections because the chocolate looks unappealing and results in large sales losses for the confectionery industry (Afoakwa et al., 2009a, Aguilera et al., 2004). Sugar bloom probably results from sugar which is dissolved by moisture and then recrystallizes subsequently on the chocolate surface. On the other hand, visible fat crystals on the surface of chocolate lead to fat bloom.

The mechanism leading to fat blooming is not fully understood yet (for example: Aguilera et al., 2004, Lonchampt and Hartel, 2004, McCarthy et al., 2003). Poor tempering and solidification of the chocolate, incompatible fats, unfavorable storage conditions such as temperature fluctuations or abrasion are possible factors which induce chocolate fat bloom (McCarthy et al., 2003, Tisoncik, 2013). Furthermore, it is likely that different mechanisms might lead to chocolate bloom[17] (for example: Hartel, 1999, Kinta and Hatta, 2005, Lonchampt and Hartel, 2004). In general, it is induced by uncontrolled crystallization of fat molecules on the chocolate.

Polymorphic transformation: Bloom might be induced by poor tempering or unfavorable storage conditions as a result of transformation of less stable crystal configurations into more stable ones. This is promoted by either improper crystallization during manufacturing or unfavorable storage temperatures which are higher than melting temperature. Poor tempering leads to crystallization into less stable crystal forms such as form IV, which melt at room temperature and later transform in an uncontrolled way

---

[17]The bloom formation mechanism likely depends on the type of chocolate product with or without filling, storage conditions and production process, thereby mainly crystallization method.

in higher crystalline forms to energetically more favorable configurations such as form V or VI. This can be seen on the surface as bloom according to Cebula and Ziegleder (1993). Afoakwa et al. (2009b) found that unstable crystals in under-tempered dark chocolate products formed into a few new but large agglomerates which are linked by solid bridges. The polymorphic form β (VI) has a strong growth tendency and fat bloom develops at the surface and the interior of the chocolate. There is a time based connection between bloom formation and the polymorphic transition (G. Ziegleder, 1994). The polymorphic form of cocoa butter of bloomed chocolate was analyzed by Cebula and Ziegleder (1993) with X-ray diffraction (XRD), differential scanning calorimetry (DSC) and nuclear magnetic resonance (NMR). They describe that bloom corresponds to different polymorphic forms and found out that there is an enhanced resistance at reduced temperatures (T = 18 °C) and with the introduction of butterfat. van Malssen et al. (1996) found that form V can rearrange into form VI via solid state transformation. Secondly, at good tempering but elevated temperature (larger than 25 °C) and changing temperature the cocoa butter transforms from form V into form VI. Another mechanism of bloom formation is where triglycerides of different types are allowed to mix. This results in a distortion of phase behavior and recrystallization of crystals with rather large and jagged morphology. The XRD data of dark chocolate at elevated temperature (T = 23 °C) indicate form VI whereas at lower temperature (smaller than 18 °C) and with the addition of 2 % milk fat form V is observed. Furthermore, the samples at form VI (dark chocolate at T = 23 °C) also show higher melting temperatures and bloom formation. This can but must not be a connection because also samples at form V show higher melting temperatures. Milk fat has the ability to hinder form V to form VI transition. Thus, fat bloom can be successfully inhibited by storage at T = 18 °C or below and addition of 2 % milk fat according to Cebula and Ziegleder (1993). Cocoa butter is able to dissolve up to 30 % milk fat without affecting polymorphism due to a low solid content within milk fat. In general, certain liquid oils accelerate the rate of transformation of form V to VI which is explained with co-crystallization of triglycerides. In contrast, certain emulsifiers retard the transformation. The transformation occurs through the solid state. The presence of an emulsifier can block the propagation of the transition from one crystal domain to the other and thus retards the transition from form V to form VI. It seems like milk fat acts similar to an emulsifier with a mechanism of retardation. Milk fat triglycerides are highly compatible with cocoa butter triglyceride and do not show phase separation at the concentration used in the study of Cebula and Ziegleder (1993). Lohman and Hartel (1994) found that addition of the high melting fractions (25 °C to 15 °C) of milk fat, which increased the hardness of the chocolate product, and the 10 °C fraction of milk fat, which was similar to low melting fractions in terms of solid fat content and hardness, inhibited bloom whereas the low melting fractions (0 °C and 5 °C) lead to acceleration of bloom. Frazier and Hartel

(2012) found that fat migrated from cookies can inhibit bloom possibly due to disruption of cocoa butter crystallization upon cooling. However, they only tested bloom behavior of chocolate chips baked in cookies after 10 days of storage at 20 °C and not for longer time periods. Within their study they observed that the bloom degree on the chocolate chips was similar for equal amount of fat migration of palm oil shortenings and olive oil added to the cookie dough into the chocolate. Although, the total oil concentration of the cookie dough differed. They also investigated fat bloom formation of chocolate chips in samples with sand mixed with oil instead of cookie dough. Thereby, they found a maximum of bloom formation at 4 % and minimum at 8 % palm oil although migration increased with increasing oil content.

Bricknell and Hartel (1998) investigated bloomed chocolate[18] samples with X-ray spectroscopy and colorimeter. They found that form VI rapidly formed after a lag phase which took up to 10 days. Polymorphic transformation as well as visual bloom appearance decreased in samples with addition of high-melting milk fat. Polymorphic transformation into form VI was present even without any visual bloom formation in samples with amorphous sucrose whereas the ones with crystalline sucrose appeared bloomed. Thus, they conclude that lack of bloom formation is caused by differences in microstructure of the samples and not the polymorphic phase change. Wang et al. (2010) confirmed that polymorphic phase change might rather be a result than the cause of bloom formation. They came to their conclusion because of lack of polymorphic transformation within the chocolate but only at the surface.

Lipid migration and phase separation: Alternatively to polymorphic transformation, Wang et al. (2010) propose phase separation as bloom formation mechanism for untempered chocolate samples and accumulation of migrated fat at the surface with accompanied re-crystallization into more stable polymorphic forms. Phase separation with subsequent fat crystal growth has been also suggested by Kinta and Hatta (2005). This is supported by Aguilera et al. (2004), Altimiras et al. (2007), Ghosh et al. (2002), Hartel (1999), Ziegleder and Schwingshandl (1998) who propose that chocolate bloom formation might be related to migration of lipids to the surface with subsequent uncontrolled recrystallization.[19]   Cocoa butter crystals might dissolve in the mobile liquid portion and are by that transported to the surface where they recrystallize. Both incompatible fats in the chocolate recipe or migration of filling fats can lead to phase separation, which might result in chocolate blooming (Lonchampt and Hartel, 2004). Ziegleder and Schwingshandl (1998) suggest that fat bloom in filled chocolate products is caused by

---

[18]Blooming of chocolate was induced by storage under cycled temperature conditions between 19 and 29 °C.

[19]More on lipid migration in chocolate is presented in Section 2.4.2 on page 30.

fat migration and influenced by crystallization speed of chocolate. Fat migration increases at elevated temperatures (Ziegleder et al., 1996b). In contrast, recrystallization of migrated lipid molecules can be described with the Avrami equation[20] and is thus faster at lower temperatures. Therefore, an optimum temperature for fat bloom formation is present which is in between the temperatures of increased migration rates at elevated temperatures and maximum crystallization rates at lower temperatures. The exact temperature depends on the chocolate product. G. Ziegleder (1994) investigated the composition of triglycerides of fat bloom from dark chocolate with high performance liquid chromatography (HPLC) and the melting behavior with DSC. Fat bloom on dark chocolate contained a high ratio of POP, POSt and StOSt. Whereas, the fat bloom of nut containing products are enriched in triolein (OOO). Two different mechanisms of fat blooming are proposed in that study: phase transition of cocoa butter (discussed earlier in this section) and fractionation of fat phase. Additionally to polymorphic transformation, the triglycerides of cocoa butter also tend to separate in mixtures with oils or do self fractionation which can be seen as an inducer for fat bloom in filled chocolates according to the study of G. Ziegleder (1994). The symmetric arrangement of cocoa butter could favor the formation of crystals in cocoa butter and tend to form mixed crystals. Cocoa butter from west Africa is usually high in symmetric components (80 %) and is quite hard compared to soft cocoa butter e.g. from Bahia with a lower content of symmetric components (79 %). A high content of symmetric components are observed in fat bloom on dark chocolate. This leads to partial fractionation and due to that multiple non-saturated triglycerides (StOO, POO) are decreasing (POP and POSt are increasing; StOSt is decreasing). In contrast to older assumptions, POP, POSt/StOSt are compatible to build mixed crystals according to G. Ziegleder (1994). Mixed crystals with a high content of POP show a strong growth tendency. As non cocoa butter fats usually have less POSt they are also usually less susceptible to fat bloom. Thus, characteristic POSt/POP amounts probably have a strong influence on fat blooming. The HPLC studies show that the diglyceride concentration of alcohol containing pralines without a sugar coating increases because alcohol migrates through the chocolate and evaporates at the surface and probably the alcohol carries polar diglyceride while diffusing to the surface. Milk free pralines with nut and nougat filling are very susceptible to fat bloom and form mixed crystals of cocoa butter and hazelnut oil. There is no evidence for short chained triglycerides of the milk fat in fat bloom. It seems like triolein, POP, POSt and StOSt are easily combined in mixed crystals and as if the four main components can easily substitute each other. Often it happens that just chocolate with a low content of hazelnut oil shows fat migration because a higher level of oil leads to a liquid fat phase so that crystallization is impeded (G. Ziegleder, 1994).

---

[20]See Section 2.2.2 about kinetics of cocoa butter crystallization on page 8 for further information on crystallization and the Avrami equation.

Lipid migration: James and Smith (2009) imaged and analyzed the surface of fresh and bloomed chocolates with X-ray photo-electron spectroscopy, cryo-scanning and environmental electron microscopy. They found that fat crystals of chocolate fat bloom appears as blade like structures, whereas sugar bloom consists of round shapes. They propose that fat blades forming chocolate bloom are extruded to the surface. Hartel (1999) summarizes that crystals at the surface start to develop nearby small cracks and crevices present at the surface. These structural defects depend on the cooling procedure according to him. Furthermore, James and Smith (2009) found that bloom on poorly tempered chocolates consist of a mixture of sugar and fat, whereas bloom on well tempered samples only comprised of cocoa butter fat. This indicates the presence of different bloom mechanisms. James and Smith (2009) proposition of extruded fats to the surface are strengthened by the findings from Smith and Dahlman (2005). They found droplets which are softer than the chocolate surface around deep holes at the surface. These structures increased in size with storage time and over time fat bloom developed. The number density of drops correlated with the degree of fat bloom formation. Thus, based on their results they propose a two-step bloom formation process involving initial drop formation out of liquid or semi-liquid oil from which bloom crystals nucleated and grew. They hypothesized that the drops might originated from lipids transported through the pipe like structures through the chocolate (possibly even from the filling) towards the surface. In contrast to that, Rousseau (2006) who imaged the surface of milk chocolate did not observe any changes of the pores themselves and thus suggest that liquid lipid migration through these pores is not directly linked to fat bloom formation. However, he observed crystal growth in the vicinity of these pores at the surface.

Lipid migration and polymorphic transformation: Hodge and Rousseau (2002) propose a mechanism involving both lipid migration and polymorphic transformation. Hodge and Rousseau (2002) visualized the surface of fresh and bloomed milk chocolate with atomic force microscopy. They induced chocolate bloom by temperature cycles between 20 and 32, 33 as well as 34 °C. Visual appearance of bloom were different for the three set-ups with most significant bloom attributes with large, jutting crystals within smooth areas for the highest maximum temperature of 34 °C. Sample surfaces which were exposed to lower maximum temperatures of 32 °C were more homogeneous and less rough without any visible bloom formation even after three cycles. Furthermore, the polymorphic configuration of the latter samples did not change significantly over the observation period. In contrast, samples which were exposed to temperatures of up to 34 °C showed increased amount of polymorph VI. Kinta and Hatta (2006) confirmed the findings from Hodge and Rousseau (2002). They also investigated bloom formation due to partial melting induced by storage under temperature cycles between 20 and 32 °C at 12 hours intervals. They propose the following formation mechanism based on Hodge and Rousseau (2002). Increase in temperature leads to partial melting and transformation from form V into form VI which is increased due to subsequent cooling. This again induces contraction and might result in liquid fat being pulled into the chocolate sample. Melting might also induce a pumping effect of liquid cocoa butter which is pushed to the surface because of volume expansion of the cocoa butter phase. Higher melting crystals might dissolve in the mobile phase and are transported to the surface. Repetition of temperature change causes accumulation of high melting fat fractions at the surface which appears as fat bloom according to Hodge and Rousseau (2002), Kinta and Hatta (2006).

2.4 Mass transport phenomena

As chocolate is a multicomponent porous material, first a theoretical background of mass transport mechanisms in porous materials in general is presented. Afterwards, studies investigating lipid migration in crystalline fat suspensions such as cocoa butter and chocolate are introduced. Transport phenomena have a key role in food technology, and thus not only for chocolate, according to Aguilera (2005).

2.4.1 Possible mass transport mechanisms

In porous materials transport processes can be divided into capillary liquid transport and non-capillary transport according to Brakel and Heertjes (1977) (Figure 2.6). Thereby, capillary driven transport is dependent on the geometry of the menisci between liquid and vapor. Capillary transport phenomena are e.g. wetting, capillary rise and infiltration. On the other hand, non-capillary transport mechanisms are molecular diffusion, viscous flow and dispersion. Viscous flow is driven by a pressure gradient, which can have different origins such as externally applied pressure or capillary pressure. Thereby, the type of transport mechanism is dependent on the material structure (especially in the range of 100 microns or less) (Aguilera, 2005).

2.4.1.1 Non-capillary transport

Molecular diffusion in unconfined pathways can be described with Fick's law which says that the driving force is a concentration gradient Δc_A (Equation 2.4). Thus, the diffusive flux $\dot{J}_A$ is:

$$\dot{J}_A = -D_{eff} \cdot \frac{\Delta c_A}{\Delta x} \tag{2.4}$$

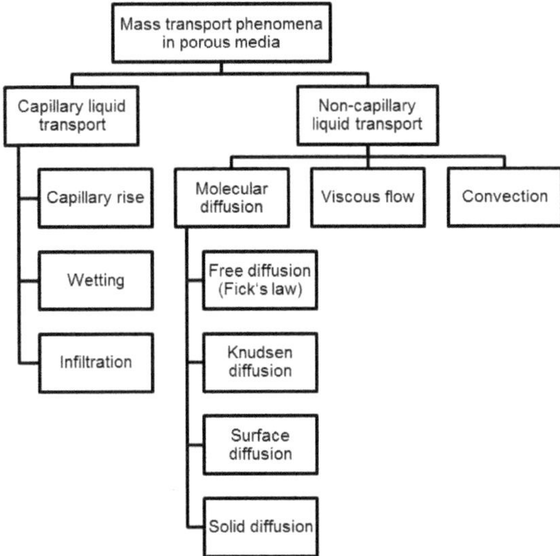

FIGURE 2.6: Categorization of transport mechanisms in porous materials based on Brakel and Heertjes (1977).

with the concentration gradient of component 'A' Δc_A over a distance Δx and the effective diffusion coefficient

$$D_{eff} = D \cdot Q \tag{2.5}$$

consisting of the diffusion coefficient D and the diffusibility Q which depends on the porespace structure (Brakel and Heertjes, 1977).

Another mechanism is convection (also denoted as mechanical dispersion or mechanical diffusion). This mechanism is a collective movement of molecules within fluids. It is caused by flow due to streamlines such as mixing. It is composed of a diffusive term and advection which is mass transport with the bulk flow (Brakel and Heertjes, 1977).

2.4.1.2 Viscous flow

Viscous flow occurs when inertia cannot be neglected (Brakel and Heertjes, 1977). Darcy describe the flow in porous media which depends on a pressure gradient Δp (Equation 2.6) (Darcy, 1856)[21]. Darcy's equation is in the same form as the law by Poiseuille for laminar flow.

$$q_{Flow} = K \cdot \frac{\Delta p}{L \cdot \eta} \tag{2.6}$$

with the constant K which describes the characteristics of the porous structure and the viscosity of the liquid η, the length L and pressure gradient Δp.

---

[21] A height difference leads to the pressure gradient in the original work of Darcy (1856).

The Knudsen number (Equation 2.7) indicates the flow regime with viscous flow for values smaller than $Kn < 0.01$, slip flow between $0.01 < Kn < 0.1$, transition flow at $0.1 < Kn < 10$ and finally Knudsen's free molecular flow with values of $Kn > 10$ for a gas flow (Ziarani and Aguilera, 2012). The Knudsen number Kn describes the ratio of the mean free path λ_{Kn} to a characteristic length, such as the channel width, e.g. $2 \cdot r_0$.

$$Kn = \frac{\lambda_{Kn}}{2 \cdot r_0} \tag{2.7}$$

Bridgman (1923) proposed to use the lattice spacing as the mean free path for liquids.

2.4.1.3 Capillary liquid transport

Capillary motion is driven by a pressure gradient which is caused by surface tension of a liquid. The pressure difference Δp is according to the Young-Laplace equation (Equation 2.8)) (Brakel and Heertjes, 1977, Young, 1805):

$$\Delta p = \gamma \cdot \left(\frac{1}{R_1} + \frac{1}{R_2} \right) = \frac{2 \cdot \gamma \cdot cos(\theta)}{r_l} \tag{2.8}$$

with the surface tension of the liquid γ and the geometry of the meniscus described with R_1 and R_2 or alternatively with the contact angle θ and r_l which is the part of the capillary radius where the liquid meniscus acts on. The smaller the space in which the liquid transport occurs, the higher the pressure. Thus, capillary pressure is only relevant for small openings. According to Brakel and Heertjes (1977) the pressure difference can be usually neglected when $(1/R_1 + 1/R_2) < 1/100 \ nm^{-1}$. The geometry of the meniscus is determined by the liquid and solid material properties and the geometry of the pores (Brakel and Heertjes, 1977). Dependent on the orientation of capillary transport capillary rise, which is an upwards motion, wetting, which is usually on a horizontal plane, and infiltration, which is downwards directed, can be distinguished (Brakel and Heertjes, 1977).

Gravity forces do not play a major role in transport over some millimeters through pores within the nanometer range and only effect transport rates at capillary sizes of some micrometers (Grüner et al., 2016, 2009). The equilibrium height h_{max} (Equation 2.9, Fries and Dreyer (2008)), at which gravity balances the capillary pressure, is in the range of several hundred meters up to kilometers in nanometer capillaries.

$$h_{max} = \frac{2 \cdot \gamma \cdot cos(\theta)}{r \cdot \rho \cdot g \cdot sin(\psi)} \tag{2.9}$$

Grüner et al. (2016, 2009) investigated transport of hydrocarbons through porous glass. The pores in the glass are quiet uniform and large and are connected by narrow throats, which limits the transport through the pore spaces. The tortuosity of the material is in between $\tau = 3.4$ and $\tau = 4.2$ and depends on shape, size distribution and topology of the pores (Lin et al., 1992). Grüner et al. (2016, 2009) describe migration based on Darcy's equation with the Laplace pressure as the driving force and neglect gravity. The equation can be solved with a Lucas Washburn approach according to Lucas (1918), Washburn (1921) and the radius r, surface tension γ, contact angle θ, viscosity η and time t (Equation 2.10).

$$h = \sqrt{\frac{r \cdot \gamma \cdot cos(\theta)}{2\eta}} \cdot t \tag{2.10}$$

A combination of Equation 2.6 and Equation 2.10 leads to Equation 2.11 presented by Grüner et al. (2016) who included the imbibition ability of the matrix material (Equation 2.13).

$$h = \sqrt{\frac{\gamma \cdot cos(\theta)}{\phi_i \cdot 2\eta}} \cdot \Gamma \cdot \sqrt{t} \tag{2.11}$$

The hydraulic permeability of the porous glass matrix K can be determined with Equation 2.12 with the radius r and hydraulic radius r_h, the tortuosity τ and volume porosity of the material ϕ_0 (Debye and Cleland, 1959, Grüner et al., 2016, Lin et al., 1992).

$$K = \frac{r_h^4 \phi_0}{8r^2\tau} \tag{2.12}$$

Grüner et al. (2016, 2009) investigated in how far macroscopic properties such as surface tension, wetting and viscosity are applicable in confined space of a few nanometers. Additionally, Grüner et al. (2016) discuss the limitations of classic hydrodynamics and continuum descriptions due to enhanced wall effects. They found no significant difference in imbibition ability Γ (Equation 2.13) for hydrocarbons with a molecular size from n-$C_{10}H_{22}$ to a size of n-$C_{60}H_{122}$ in case of porous glass with pore spaces with a mean diameter of 3.5 nm and 5 nm except of a slightly higher value for the largest tested molecule n-$C_{60}H_{122}$ through glass with the smaller mean diameter of 3.5 nm. They explain the deviation with possible alignment or configurational changes of the molecule due to the strong confinement.

$$\Gamma = \frac{r_h^2}{r_0} \cdot \sqrt{\frac{\phi_0}{\tau \cdot r_l}} \tag{2.13}$$

The imbibition ability from gravimetric measurements for the porous glass with nominal pore sizes of 3.5 nm (nitrogen sorption measurements revealed a mean pore size of 3.4 nm) is $\Gamma \approx 125 \pm 3$ x $10^{-7}\sqrt{m}$ and for 5.0 nm it is $\Gamma \approx 180 \pm 2$ x $10^{-7}\sqrt{m}$ (Grüner et al., 2016). The branched squalene hexamethyltetracosane showed slightly less imbibition ability into the porous Vycor glass. They explain this phenomenon with sticking layers at the pore wall. Imbibition abilities of the 3.5 nm porous glass determined based on detection of migration front from optical measurements are in the range of $\Gamma \approx 92.6 \pm 6.4$ x $10^{-7}\sqrt{m}$ for C14 and C12 at relative humidities between 24 % and 50 % for the lower bound and $\Gamma \approx 120.6 \pm 2.4$ x $10^{-7}\sqrt{m}$ for the upper bound. Grüner et al. (2016) conclude that the transport of hydrocarbons through the porous glass is capillarity driven and can be described with the classical Lucas-Washburn equation (Equation 2.10) used for spontaneous imbibition in porous media. The bulk fluid properties can be used for the transport description within the nanometer confined spaces of the porous glass. Stukan et al. (2010) propose to use Lucas-Washburn equation with a dynamic contact angle. Furthermore, Grüner et al. (2016) observed sticking of flat-lying hydrocarbons and a monolayer of water adsorbed at the pore wall, which narrow the effective pore size in which transport occurs. Stukan et al. (2010) found that an interaction of the fluid in the nanometer confined spaces with the wall can lead to a local increase in viscosity but should not effect the viscosity in the central part of the pore space. Furthermore, roughness of the confined spaces generally slows down the meniscus velocity during spontaneous imbibition and inhibits wetting (Stukan et al., 2010). This decreases the hydrodynamic radius of the capillary.

2.4.2   Lipid migration in crystalline fat suspensions

Lipid migration in chocolate products is a common problem in confectionery industry (e.g. Ghosh et al., 2002, Ziegler, 2007). Migration rates in general are positively dependent on the driving force and negatively related to resistance. Thus, to reduce lipid migration either driving forces have to be eliminated or the resistance has to be increased (Ziegler, 2007). Densely packed structures might inhibit migration whereas in contrast imperfections such as cracks and crevices, which might originate from improper solidification, might enhance migration by being direct transport pathways (Hartel, 1999). But the mechanism for lipid migration is not fully understood yet (Ghosh et al., 2002, Van der Weeën et al., 2013). It is likely that diffusive mixing and capillary transport through micro- and nanopores both contribute to lipid migration in chocolate (Aguilera et al., 2004, Rousseau and Smith, 2008). Aguilera (2005), Aguilera et al. (2004) specify that migration in food materials is probably mostly influenced by structural elements which are smaller than 100 µm. The migration rate depends on the structure[22] and composition of the chocolate product (Ghosh et al., 2002). Ghosh et al. (2002) claim that migration occurs as long as there is a difference in the chemical potential and the

---

[22]Chocolate structure is presented in more detail in Section 2.2.2.3 on page 14.

pathway in the continuous lipid phase is mainly through the liquid part. Thus, an increase in liquid fat content would lead to more migration. Ghosh et al. (2002) also point out the importance of structure on migration rates in their review. Generally, a dense structure results in reduced migration. Furthermore, it seems like the mechanism depends on other aspects such as temperature and type of chocolate product, for example kind of filling (Dahlenborg et al., 2011, 2012, Rousseau and Smith, 2008). Diffusion and capillary rise are both widely discussed as possible transport mechanisms in chocolate.

2.4.2.1 Diffusion

Lipid migration in chocolate is often described as a diffusion based mechanism. A possible driving force for diffusion could be a concentration gradient of triglycerides (Ziegleder et al., 1996a, Ziegler, 2007).

For migration in chocolate the diffusion equation can be approximated by mass gain with the mass m_t at time t and mass at saturation m_s, the contact area A and volume V as well as the effective diffusion coefficient D_{eff} (Equation 2.14) (Ziegleder et al., 1996a).

$$\frac{m_t}{m_s} = \frac{A \cdot \sqrt{D_{eff} \cdot t}}{V} \tag{2.14}$$

Change of structure due to swelling or eutectic effects is neglected and contact between the phases is assumed in this equation. However, eutectic effects are present in chocolate systems and lead to over proportional softening of the chocolate product[23].

Ziegleder et al. (1996a,b), Ziegleder and Schwingshandl (1998) applied the diffusion equation to describe migration from a nougat filling into milk chocolate which they traced back with high pressure liquid chromatography (HPLC). Linear increase of mass with the square root of time was observed for the first 25 days after which migration slowed down and reached a saturation value after about 100 days at 26 °C. In general, migration is slowed down with decrease in storage temperature (e.g. Smith et al. (2007), Talbot et al. (2006)). The driving force for migration is a concentration gradient of low melting triglycerides, which are liquid at room temperature, such as OOO and LOO from the nougat filling into the chocolate according to Ziegleder et al. (1996a). As a counterpart, Ziegleder et al. (1996a) observed migration of POP and PLSt from the chocolate into the nougat filling. However, counter migration is less than migration from nougat filling into chocolate. Furthermore, concentration of POSt is basically constant probably due to very small concentration gradient and crystallinity within measurement conditions.

---

[23]The eutectic effect is introduced in Section 2.2.2.6 on page 18.

Khan and Rousseau (2006) found that the migration mechanism might change with storage temperature. They analyzed migration of hazelnut oil into dark chocolate at a temperature of 26 °C and found that it is more likely non-Fickian diffusion whereas at 11 °C migration follows a pseudo-Fickian behavior and at 20 °C it resembles Fickian one. Effective diffusion coefficients of chocolate and cocoa butter can be extracted from experimental data of migration via Equation 2.14 (Figure 2.7). The diffusion coefficient is dependent on the storage temperature and tempering of the sample. Diffusion coefficients between $D_{eff} = 10^{-15} m^2/s$ to $D_{eff} = 10^{-10} m^2/s$ are reported in literature (Figure 2.7). Maleky and Marangoni (2011a), Marty et al. (2005) visualized migration from an oil with palmfat filling into tempered and untempered cocoa butter with a flatbed scanner. Maleky and Marangoni (2011b) found that the effective diffusion coefficient is lower in case of immersed cocoa butter sample into the filling (system B) than when the cocoa butter sample is solely aligned with the filling (system A). Furthermore, migration is faster in samples which have been statically crystallized compared to sheared and oriented crystallized samples (Maleky and Marangoni, 2011a). They determined effective diffusion coefficients from their experimentally measured migration heights per time. Figure 2.7 summarizes effective diffusion rates reported in literature. Khan and Rousseau (2006) measured the mass uptake of a hazelnut oil and icing sugar filling into dark chocolate. Ziegleder et al. (1996a) investigated migration from triolein into chocolate. Walter and Cornillon (2002) used NMR technique to analyze migration of lauric acid into dark chocolate. Choi et al. (2007) measured migration from peanut butter into chocolate with different mean particle sizes. Miquel analyzed the migration of a filling of hazelnut oil and sugar into dark chocolate with different tempering degrees.

Galdámez et al. (2009) propose a predictive model for oil migration in chocolate in terms of diffusion considering molecular diffusivity and the inner microstructure. They include moving boundaries to account for swelling of the chocolate matrix and tortuosity which accounts for the embedded particles and varies with oil concentration. The driving force for migration in their model is a concentration gradient in liquid lipids within the liquid parts of chocolate.

However, according to Ziegler (2007) the migration process in chocolate is a complex phenomena which possibly involves both diffusion and phase equilibrium. Oil migration from e.g. liquid nut fillings leads to dissolution of solid fat of the chocolate into which migration takes place and thus disturbs phase equilibrium. This again increase the driving force for diffusion. Thus the amount of liquid phase is of high importance for oil migration according to Ziegler (2007). Lipid migration in chocolate is complex due to heterogeneities in structure and composition which may both vary with time and thus

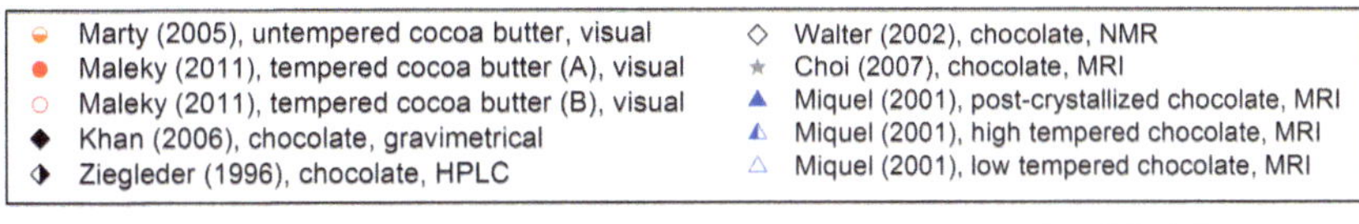

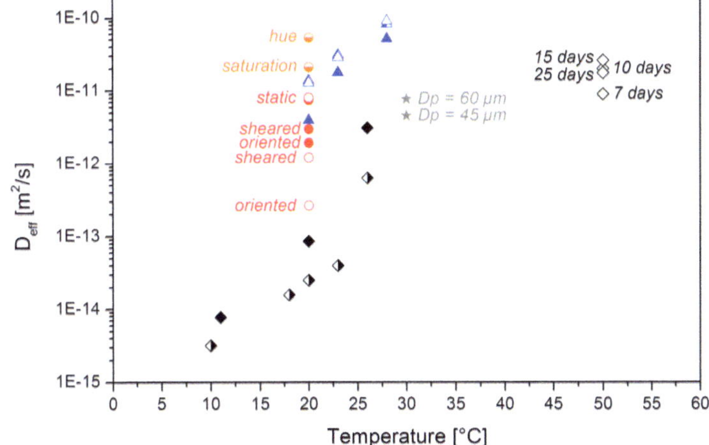

FIGURE 2.7: Effective diffusion coefficients reported in literature in dependence of storage temperature (Choi et al., 2007, Khan and Rousseau, 2006, Maleky and Marangoni, 2011a, Marty et al., 2005, Walter and Cornillon, 2002, Ziegleder et al., 1996a).

lipid migration in chocolate cannot be solely modeled with the simple Fickian diffusion equation (Choi et al., 2007, Green and Rousseau, 2014).

2.4.2.2 Capillary flow and pressure gradient

Besides diffusion, capillary flow of the liquid fat portion through interconnected pores and passages of the particular structure of chocolate is a possible migration mechanism in chocolate. Afoakwa et al. (2009a) speculated that capillary transport in chocolate might occur in the spaces between the particle network filled with both liquid and solid fat or through capillaries between fat crystals filled with liquid fat. Aguilera et al. (2004) summarize data on the influence of microstructure and temperature as well as kinetic results to come to the conclusion that capillary flow probably significantly contributes to migration in chocolate. They modeled capillary flow with the Lucas-Washburn equation (see Equation 2.10, Section 2.4.1.3 on page 28). However, due to lack of detailed data on microstructure of chocolate and properties of migrating lipids such as surface tension, viscosity and contact angle Aguilera et al. (2004) cannot verify unambiguously their hypothesis.

Van der Weeën et al. (2013) propose a cellular automata model to describe oil migration in chocolate which means that time and spatial as well as state domains are discrete. The model is calibrated with migration experiments. Detailed data on structure and physiochemical properties of the system would allow a more precisely determination of model constants. The advantage of the model by Van der Weeën et al. (2013) is that each cell can be given different states to contribute for heterogeneity within the sample. The model allows to include capillaries in combination with a diffusion gradient and structural information such as from tomography studies. Thus, combined mechanism of diffusion and capillary flow can be taken into account.

Dahlenborg et al. (2011, 2012) suggest that convective flow through pores and cracks is the main mechanism for lipid migration in chocolate stored at constant temperature[24]. Thereby, they observed formation of oil protrusions from which fat crystals nucleated and grew. In case of cycled temperatures they hypothesized convective flow in both directions due to a pushing effect of lipids through the pores to the surface and backwards. In accordance to that, Loisel et al. (1997) also propose that liquid fractions of cocoa butter might be pushed to the surface due to pressure gradient during temperature variation which leads to contraction (at solidification) and expansion (when melting) of cocoa butter.

2.4.2.3 Qualitatively and quantitatively investigation

Miquel et al. (2001), Choi et al. (2007) and Guiheneuf et al. (1997) used magnetic resonance imaging (MRI) to analyze quantitatively and qualitatively the kinetics of low melting filling fats such as hazelnut oil and peanut butter into chocolate without sample destruction. Liquid lipid parts are visualized with MRI and thus enables to detect spatial differences of liquid lipid content in a sample and therefore migration from high liquid lipid fillings into chocolate with lower liquid lipid content. They found that migration is higher at elevated temperature[25]. Different tempering (untempered, tempered and post crystallization) had no effect on migration rates according to Miquel et al. (2001). In addition, Miquel et al. (2001) modeled migration with diffusion based equation, whereas Guiheneuf et al. (1997) propose a combined migration mechanism of diffusion and capillary flow. Choi et al. (2007) measured faster migration with larger particles (average particle size of 60 µm compared to 45 µm). This is consistent with both diffusion and capillary pressure driven migration at small times.

As another technique, Marty et al. (2009) imaged and quantified migration of a soft fat stained with nile red, a lipid soluble dye, into tempered cocoa butter with a flat

---

[24]They observed surface changes on filled white chocolate pralines with microscope techniques.
[25]28 °C compared to 19, 20 and 23 °C.

bed scanner. They found that oil migration and intensity of dye is linearly related within the first 28 days and that the amount of migrated dye was highly affected by the matrix structure. Smith et al. (2007) analyzed migration of hazelnut oil into cocoa butter by investigation of different layers with gas liquid chromatography which requires sample destruction. As an advantage, this technique enabled them to evaluate other properties such as solid fat content and polymorphism besides amount of migrated filling. Furthermore, migration can also be measured with a gravimetric method (Svanberg et al., 2012). Alternatively, Walter and Cornillon (2002) analyzed the kinetics of lipid migration in chocolate at 50 °C for accelerated migration with lauric acid as migrant. They used magnetic resonance and differential scanning calorimetry (DSC) techniques. Migration started after 6 days of induction time. A possible explanation is that migration might only take place when a particular internal structure of chocolate is reached. Thus, prior to migration of lauric acid into cocoa butter, phase separation of the triglycerides and accumulation of mobile cocoa butter fractions at the contact area of chocolate and lauric acid might take place. They found that even after start of migration, the rate changes over time with an increase for the first 15 days and slight decrease afterwards. This might be caused by capillary nature of fat migration and probably involves both diffusion and capillary rise (Walter and Cornillon, 2002).

2.4.3 Impact of structure on lipid migration

Maleky and Marangoni (2011a), Maleky et al. (2012) investigated the influence of structure of cocoa butter on migration of stained triglycerides as well as filling fats by magnetic resonance imaging (MRI). They induced different crystal network structures by alteration of the solidification process under shear and static crystallization. All in all, the finer and more homogeneous the structure with lower permeability and thus higher tortuosity, the slower migration. Thereby, size and distribution of crystal network as well as its geometry both impact migration rates. The results from migration of filling fats observed with MRI are comparable to the ones obtained by visual inspection of stained triglyceride migration (Maleky and Marangoni, 2011a, Maleky et al., 2012).

Regarding addition of a second component, Altimiras et al. (2007) found that cocoa butter with small particles is more prone to lipid migration than samples with larger particle sizes.[26] In contrast, Choi et al. (2005) measured an increased migration rate with increasing particle size. However, the amount of migration was not significantly influenced by the particle size of the chocolate. They concluded that the higher the

---

[26]In the publication sand with an average sizes of 118.60, 55.24 and 4.60 µm was used as particular phase and storage temperature was 30 °C.

porosity, which is increased at larger particle size according to them, the easier is fast oil migration but the total amount is unaffected.[27]

Furthermore, Marty and Marangoni (2009) visually observed a dependence of migration rates on cocoa butter origin. Thereby, the higher the unsaturated triglyceride content and oleic acid the faster the lipid transport. The impact of cocoa butter origin is reduced in case of tempering. In general, lipid migration in tempered cocoa butter were found to be 10 to 50 times slower than without tempering. Besides, the lag phase, which describes the time before significant migration starts, is longer in case of tempered samples (Marty and Marangoni, 2009). A possible explanation for the lag phase is the dynamic characteristic of the interface between filling and cocoa butter.

The role of structure on lipid migration was also discussed for other food products. For example, Bouchon et al. (2003) suggest that microstructure also highly impacts the uptake of oil stained with Sudan red into potato chips which they measured by means of a spectrophotometer and observed with confocal laser scanning microscopy.

---

[27]They measured oil migration from a peanut butter filling into chocolate with MRI technique. Particle size was 45 μm to 60 μm.

3

Materials and Methods

This chapter presents the materials and methods used. It starts with sample preparation. Then, the small angle X-ray scattering and corresponding contact angle experiments are described. The X-ray tomography experimental set-up is subsequently introduced followed by the stress simulations with means of finite element method. Afterwards, the experimental set-up to determine migration on a macroscopic level is shown. This is accompanied by the description of simulation of the migration. And finally, additional analytic tests used in the framework of the thesis are presented.

3.1 Sample preparation

Chocolate and chocolate model systems in powder, solid form and as filled products were investigated in the framework of this thesis in order to analyze different aspects of lipid migration in chocolate. Chocolate is a product composed of particles embedded in cocoa butter. The material properties depend on the composition and nature of components. Furthermore, production conditions and especially the tempering (solidification due to crystallization under controlled temperature and agitation) process significantly influence the state of cocoa butter and thus chocolate properties.

Samples were used in powder forms and as solid bars with different geometries adjusted to the experimental conditions of the different experiments. Thus, powder samples were used for small angle X-Ray scattering to ensure migration within measurement time. Conventional chocolate samples and model systems from cocoa butter or particle in fat suspensions were used for structure and surface analysis with tomography and contact angle measurements and to study migration on a macroscopic level.

3.1.1   Material for small angle X-ray scattering

The materials (cocoa and skim milk powder, sucrose and cocoa butter) used for small angle X-ray scattering were provided by the Nestlé Product Technology Center (PTC) York, UK. Natural cocoa powder with a fat content of 11 ± 1 % and medium heated skimmed milk were used as powder materials without any pretreatment. Furthermore, refiner flakes with 27 % cocoa butter and skimmed milk powder with a particle size of approximately 15 µm were produced. Additionally, sucrose powder was refined with 23 % cocoa butter to obtain cocoa butter coated flakes with a particle size of approximately 20 µm. Food retail sunflower oil (VITA D'OR, Brökelmann & Co. Oelmühle GmbH & Co. KG, Germany) with 92 wt% fat (10 wt% saturated fatty acids, 29 wt% monounsaturated, 53 wt% polyunsaturated (omega-6-linoleic acid)) was used as oil migrant for both µSAXS and contact angle experiments.

3.1.2   Chocolate model systems for contact angle measurements

Chocolate model systems out of pure cocoa butter (Nestlé PTC, UK), and a cocoa butter suspension with 33 wt% embedded particles were prepared as solid bars. As particles either cocoa powder (Bensdorp, Germany, particle size distribution of cocoa powder is depicted in Figure 3.1), skimmed milk powder (Nestlé PTC York) or sucrose (Sweet Family Nordzucker Puderzucker, Germany, particle size from 1 micron to 300 microns, particle size distribution of cocoa powder is depicted in Figure 3.1) were used. Additional cocoa butter composites with lecithin and sucrose have been prepared as follows. 34 wt% of sucrose with addition of about 0.2 wt% of lecithin was refined in a 3 roller refiner to a final particle size of $d_{50,3} = 11.17$ µm (from $d_{10,3} = 2.48$ µm to $d_{90,3} = 39.87$ µm)[1] (Figure 3.2). 35.9 wt% of milk powder with addition of about 0.2 wt% of lecithin was refined in a 3 roller refiner to a final particle size of $d_{50,3} = 21.33$ µm (from $d_{10,3} = 3.63$ µm to $d_{90,3} = 44.21$ µm).

Tempering with addition of seeds:

To ensure melting of all cocoa butter crystals prior to sample preparation, cocoa butter and chocolate mass were held at 50 °C for at least 1 hour. Subsequently, tempering was performed by addition of seeding crystals to achieve a stable chocolate sample in the desired crystal polymorphic form. Therefore, the molten mass was cooled to 31.5 °C, hold at that temperature for 15 minutes and about 1 wt% of crystal seeds (Nestlé PTC, UK) in form V were added under constant stirring for 10 more minutes. The molten mass was cooled at 3 ± 2 °C in a fridge for 90 minutes after tempering.

---

[1]Particle size distribution of final product was measured with the HELOS particle size analyzer at the Nestlé factory Wandsbek

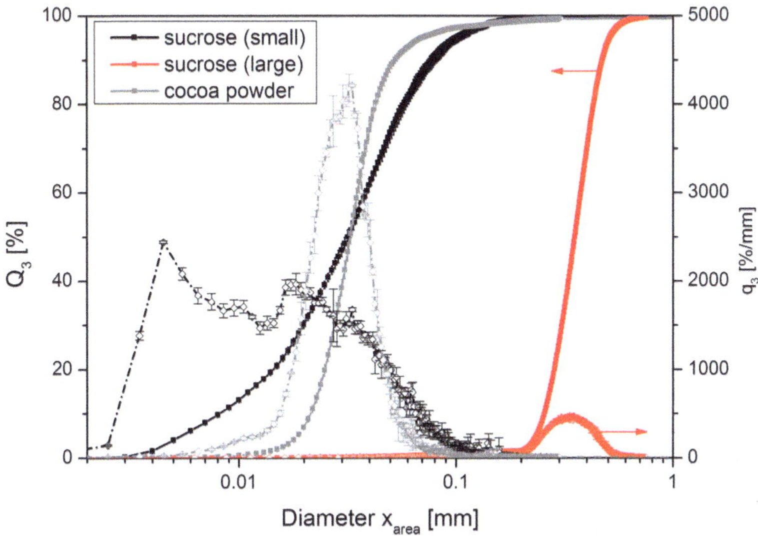

FIGURE 3.1: Particle size distribution of icing sugar (small sucrose), household sugar (large sucrose) and cocoa powder.

3.1.3 Samples for tomography experiment

Chocolate, pure cocoa butter as well as sucrose suspended in cocoa butter were imaged with tomography. Therefore, cylindrical shaped samples with a maximum diameter of about 3 mm for synchrotron based tomography and up to 5 mm for experiments with the laboratory tomography device were prepared. The chocolate masses were a standard industrial made dark and milk chocolate and were received from the Nestlé factory Wandsbek, Hamburg, Germany. The chocolate masses are directly taken from the production process after refining and conching. The dark chocolate is composed of approximately 48 wt% sucrose, 32 wt% lipid phase (30.5 wt% cocoa butter with addition of 1.5 wt% milk fat and 0.2 wt% lecithin) and 20 wt% of cocoa solids. The average particle size of the chocolate is $d_{50,3} \approx 10$ µm determined with laser diffraction. The milk chocolate is composed of approximately 43 wt% sucrose, 23 wt% skimmed milk powder, 30 wt% lipid phase (26 wt% cocoa butter with addition of 3.5 wt% milk fat and 0.5 wt% lecithin) and 4 wt% of cocoa solids. Additional cocoa butter composites with lecithin and sucrose have been prepared as described in the previous section on page 38.

Untempered samples without precrystallization were solidified directly from the molten mass at 20 °C without controlled temperature nor shear profile. The chocolate mass was

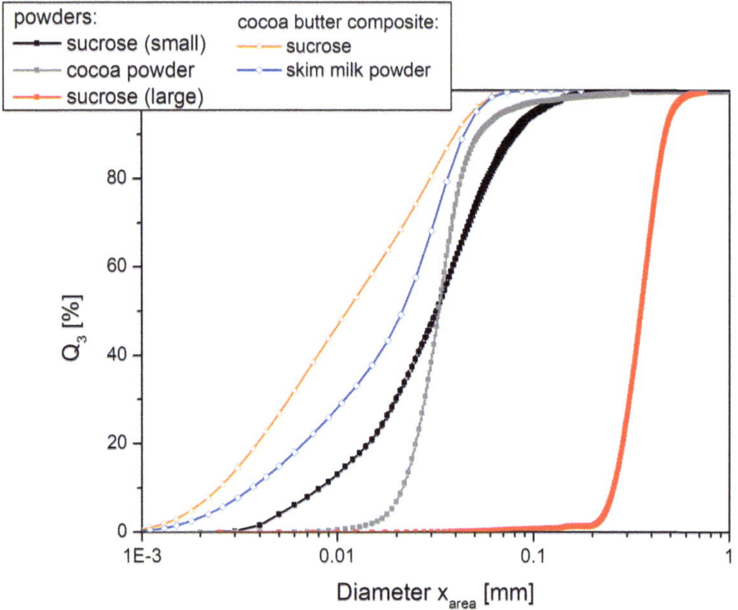

FIGURE 3.2: Particle size distribution of the particles of the cocoa butter composites.

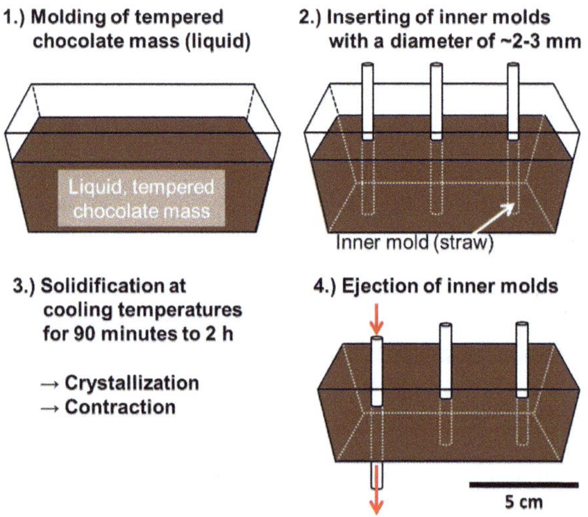

FIGURE 3.3: Preparation of cylindrical samples for tomography.

precrystallized by addition of seeds[2] or by traditional hand tempering on a marble top[3] to achieve a stable chocolate sample in the desired crystal polymorph. This leads to chocolate and model systems with the characteristic gloss and snap, which contracted in the cooling process. The molding process is depicted in Figure 3.3. The maximum diameter of the sample was restricted by the field of view, which is dependent on the X-Ray tomography set-up, thus the maximum diameter for the chosen set-up with synchrotron radiation was D = 1.8 mm x 2 = 3.6 mm. Thus, cylindrical molds with a diameter of D = 3 mm out of polypropylene were used (Figure 3.3). Samples for laboratory tomography device had a diameter of up to 5 mm. A direct filling from the molten mass into these cylindrical molds lead to incomplete filling of the molds. Therefore, the liquid chocolate mass was filled into larger molds first (Figure 3.3, step 1) and the cylindrical molds were then immediately inserted into the bed of liquid chocolate mass to ensure complete filling and wetting of the cylindrical molds (Figure 3.3, step 2). To release air entrapped during the molding process, the filled molds were put on a vibratory plate for some seconds or were tapped onto a hard base until no visible air bubbles were arising anymore. The filled mold was immediately afterwards stored at 3 ± 2 °C for 90 minutes (Figure 3.3, step 3)[4] to allow for cooling and solidification. As chocolate mass contracts during crystallization, samples could easily be demolded afterwards by gently pushing the cylindrical molds from the chocolate bed (Figure 3.3, step 4). The samples were kept and imaged inside the cylindrical molds. The samples were subsequently stored in a climate chamber at 20 °C prior to measurement.

To investigate structure change during storage at room temperature compared to elevated temperature (32 °C) an artificial capillary was molded into selected samples and imaged with standard laboratory X-ray tomography. Pure and aerated cocoa butter as well as cocoa butter suspensions with small and large sucrose particles were analyzed for that purpose. Therefore, a polymer fiber out of Nylon with a diameter of 350 μm was inserted into the liquid tempered cocoa butter mass directly after molding. The fiber was gently removed after solidification by pulling it out of the solid sample and left a cylindrical shaped capillary inside the sample. Fresh samples and samples after 9 weeks of storage were imaged. Thus, a sample of each composition was stored for 9 weeks at room temperature and another sample was stored for 3 weeks at 32 °C and additional 6 weeks at room temperature.

---

[2]Seed added tempering is described in Section 3.1.2 on page 38. Model systems were solely tempered with seeds.

[3]Hand tempering was only used for selected chocolate samples.

[4]Some tempered chocolate samples were cooled at 10 °C for 2 hours. No detectable difference in microstructure between both cooling procedures were seen.

3.1.4 Samples for macroscopic migration observation

Figure 3.4 gives an overview over the different sample set-ups used to observe migration on a macroscopic level. Four different experimental set-ups were investigated. First, migration of a high oily filling through porous glass is observed (system A, Figure 3.4). The second experimental set-up is migration of the high oily filling through cocoa butter and white chocolate (system B, Figure 3.4). The third and forth experimental set-up is migration from one layer into the other layers, whereby the layer composition is held equal (except addition of stain for visualization of migration). Thereby, all layers are either tempered or all untempered in system C (Figure 3.4). In contrast, tempering degree is varied in system D (Figure 3.4).

System B consists of only one pure cocoa butter or white chocolate layer (tempered[5] as well as untempered) on top of a soft stained filling layer (70 wt% of coconut fat Palmin®, Peter Kölln KgaA, Elmshorn, Germany, and 30 wt% of sunflower oil, VITA D'OR, Brökelmann & Co. Oelmühle GmbH & Co. KG, Germany and 0.1 g Sudan Red per 40 g fat mixture. Sudan Red (Sigma Aldrich, Germany) is an oil soluble red colored dye in powder form. The fat mixtures were stirred with a magnetic stirrer at elevated temperature to ensure a homogeneous distribution, tempered if necessary and molded into an aluminum beaker. The height of the cocoa butter layer on top were 10 mm (system B). Both layers were cooled for 90 minutes at 3 ± 2 °C and the cocoa butter layer was demolded afterwards. The cocoa butter layer was subsequently placed on top of the filling layer which was kept in the aluminum mold. The samples were stored at 20 °C in a climate chamber. In addition to pure cocoa butter samples, white chocolate ("Die Weisse" from Nestlé, Switzerland) was investigated as the top layer. Furthermore, sunflower oil was added in a concentration of 30 wt% to both pure cocoa butter and white chocolate for selected samples.

In addition, samples (system C and D, Figure 3.4) with three layers were prepared. The three layers were sequentially molded into a cylindrical plastic mold with an inner diameter of 25 mm, a wall thickness of about 1 mm and a height of around 25 mm. The samples were stored at 3 ± 2 °C in a fridge for 30 minutes after each molding step. The single layers had a height of $h = (6 \pm 1)$ mm. The samples were tapped on a table for about 3 to 10 times until no visible air bubbles rise to the surface anymore to release entrapped air. The next layer was molded on top of the prior layer after solidification for 30 minutes of the prior one. Thus, the last layer (layer 1) was stored at 5 °C for 30 minutes, the second layer for 60 minutes and the first bottom layer (layer 3) for 90 minutes in total (Figure 3.5). One layer was stained with approximately 0.1 g per layer Sudan Red. Dye concentration was kept constant to reduce any effect of chemical

---

[5]See Section 3.1.2 on page 38 for detailed description of tempering method.

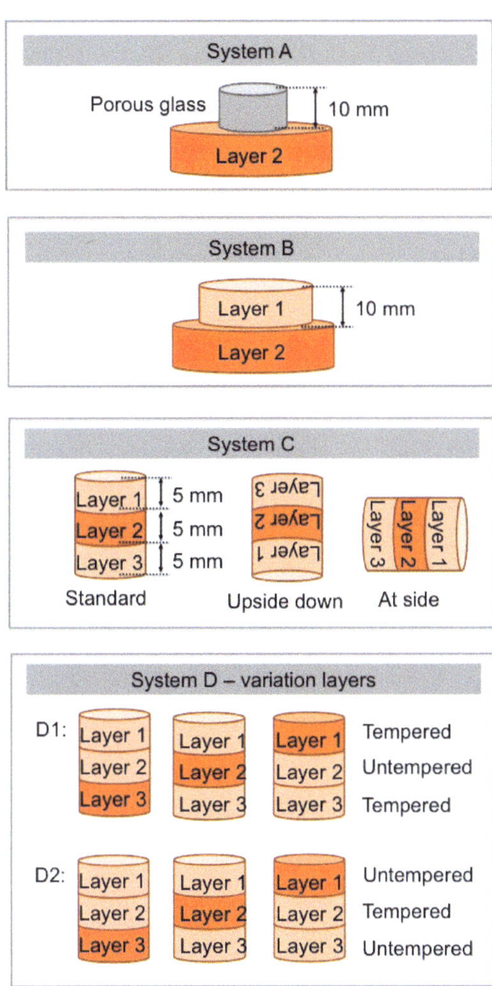

FIGURE 3.4: Experimental overview of different set-ups for macroscopic migration.

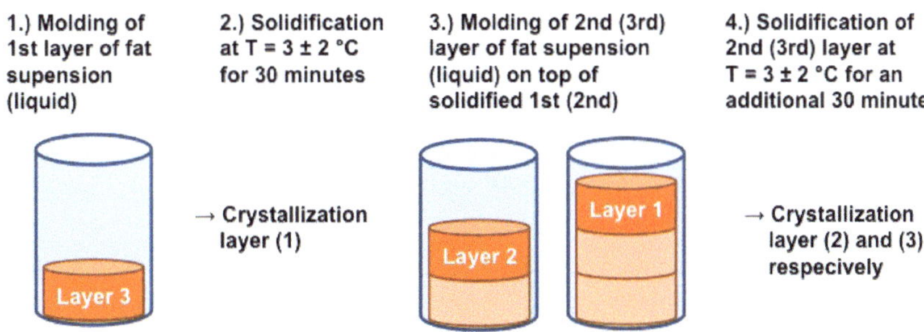

FIGURE 3.5: Experimental overview of molding process for macroscopic migration.

potential and only investigate how the structure of the samples influence the migration rate. System C was a three layer set-up with same composition in every layer, only exception is addition of Sudan Red to one of the layers (approximately 0.1 g per layer Sudan Red). Selected samples of system C type were stored upside down and at the side in addition to the standard set-up. In system D, the layers were varied in degree of precrystallization (tempered vs. untempered).

In addition to pure cocoa butter, sunflower oil (VITA D'OR, Brökelmann & Co. Oelmühle GmbH & Co. KG, Germany) or sucrose particle were added in different concentrations for selected samples with the set-up of system C. In case of sunflower oil, 5, 10, 20 and 30 wt% were added under constant agitation prior to molding. Furthermore, 1/3 per weight of sucrose (icing sugar with small particle size and household sugar with a large particle size, Figure 3.1) was added under constant agitation to the cocoa butter prior to molding for selected samples. The particle size of icing sugar ranges from $d_{10,3} = 8$ µm to $d_{90,3} = 79$ µm and an average diameter of $d_{50,3} = 32$ µm, that of cocoa powder is $d_{10,3} = 21$ µm to $d_{90,3} = 52$ µm and an average diameter of $d_{50,3} = 33$ µm and household sugar has particles from $d_{10,3} = 256$ µm to $d_{90,3} = 468$ µm and an average diameter of $d_{50,3} = 352$ µm based on $Q_3(x_{area})$ measured with the Camsizer XT, Retsch based on dynamic image analysis according to ISO 13322-2 (Figure 3.1). The sphericity of the particles is on average 0.84 for cocoa powder, 0.83 for icing sugar and 0.86 for household sugar. Additional cocoa butter composites with lecithin and sucrose have been prepared as follows. 34 wt% of sucrose with addition of about 0.2 wt% of lecithin was refined in a 3 roller refiner to a final particle size of $d_{50,3} = 11.17$ µm (from $d_{10,3} = 2.48$ µm to $d_{90,3} = 39.87$ µm)[6] (Figure 3.2).

3.2 Methods

Samples were analyzed with micro small angle X-ray scattering (µSAXS). Contact angles were measured with the sessile drop technique. Tomography imaging was performed and visual observation of macroscopic height change of the stained layer into unstained layers was carried out.

3.2.1 Small angle X-ray scattering

The experimental set-up for small angle X-ray scattering (SAXS) was also described in Reinke et al. (2015b). Figure 3.6 gives an overview of the experimental set-up. µSAXS

---

[6]Particle size distribution of final product was measured with the HELOS particle size analyzer at the Nestlé factory Wandsbek

measurements were performed at the MiNaXS/P03-beamline of the PETRA III storage ring at DESY (Deutsches Elektronen Synchrotron). A beamsize of (31 x 22) μm^2 was used with a wavelength of $\lambda = 1.090 \pm 0.005$ Å. The sample to detector distance was (3651 $\pm$ 1) mm. The 1M Pilatus detector (Dectris Ltd., Switzerland) had a pixel size of 172 x 172 μm^2. The powder samples were placed in a 1 mm thick sample cell made of Kapton foil (thickness: 76 μm). The sample volume was 0.12 ± 0.01 cm^3. Instead of single X-ray exposure at a fixed position, a vertical scan (Figure 3.6) with 61 points (step size: 100 μm) was performed for each time step to account for effects due to sample heterogeneity. The use of a small microfocus beam in a scan mode with the given experimental procedure ensures no beam damage of the sample and a good signal-to-noise ratio. A preliminary test proofed that the structure is not modified due to beam exposure for 1 s. In addition, by using a microfocused beam, truly statistical values are extracted and deviations from the mean values can be evaluated, rather than only relying on average macroscopic values. A first scan was performed before the addition of oil. Then, approximately 10 μl of oil were deposited through an upper aperture in the sample cell and a second vertical scan was performed (Figure 3.6). A vertical scan was then

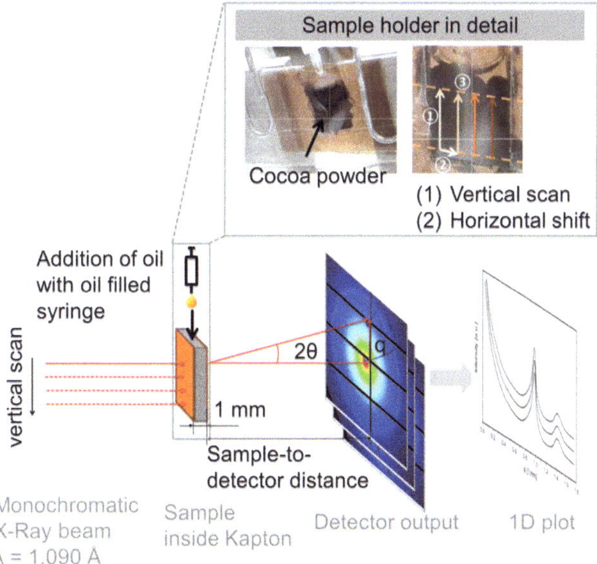

FIGURE 3.6: Experimental overview of set-up for oil migration tracking into chocolate components with small angle x-ray scattering.

repeated after 5, 10 and 30 minutes of oil addition, and, for selected samples, additional scans after 1, 2, 5 hours and 1 day were realized. To avoid beam induced effects the exposure time was set to 1 s and a horizontal shift of 500 μm was realized between every vertical scan.

The intensity I is dependent on the scattering vector q (Equation 3.1) which is a function of $2 \cdot \Theta$, which is the angle between the incident X-ray beam and the detector (Figure 3.6) and the wavelength of the X-rays λ.

$$q = \frac{4\pi sin(\Theta)}{\lambda} \tag{3.1}$$

For each vertical scan 61 scattering patterns were summed up. A mask was applied to account for the intermodular detector gaps and the beam stop (used to prevent the detector from oversaturation due to the high X-ray intensity) and standard normalization was performed. Afterwards, the 2D scattering patterns were transformed into 1D data plots by radial integration using the DPDAK software package (Benecke et al., 2014). The scattering peaks were fitted using the DPDAK software with a pseudo-Voigt profile combined with a linear background. The scattering power k is the total scattered integral intensity, which depends on the electron density difference $\Delta\rho$, the volume fraction ϕ and the irridated volume V:

$$k = \int I(q) \cdot d^3 \cdot q = (\Delta\rho)^2 \cdot \phi \cdot (1 - \phi) \cdot V \tag{3.2}$$

The equation applies for non-particulate two-phase systems (Porod, 1982). The scattering power k* (Equation 3.3) of the measured 1D data system has been calculated in Origin (OriginLab, Northampton, MA, United States) according to Equation 3.3.

$$k^* = \int_{q_{min}=0.05nm^{-1}}^{q_{max}=1.71nm^{-1}} I(q) \cdot d^3 \cdot q \tag{3.3}$$

For powders having highly ordered structures, the peak intensity was calculated with Equation 3.4.

$$k_{peak} = \int I_{fit}(q) \cdot d^3 \cdot q - \int I_{baseline}(q) \cdot d \cdot q \tag{3.4}$$

Thereby, $I_{fit}(q)$ is the fitted function of the peak intensity with the scattering vector. $I_{baseline}(q)$ is the function of the baseline, which was approximated as a line. Peak integration was performed with Origin.

The difference of the total scattering power and all peak intensities is then Equation 3.5.

$$\Delta k = k^* - \sum k_{peak} \tag{3.5}$$

3.2.2   Contact angle measurements

The contact angle was measured with the OCA 20 instrument from dataphysics, Filderstadt, Germany or with the sessile drop protocol applied to an inhouse developed setup. Therefore, a droplet with a volume of 5.3 $\pm$ 0.3 µl with the OCA 20 instrument and 0.7 $\pm$ 0.1 µl with the inhouse device was deposited on the surface of the chocolate model samples with a pipette. Images of the droplet were acquired with a highspeed camera (Serie NX-S2) connected to a microscope objective from OPTEM ZOOM 125, Qioptiq, Excelitas Technologies. Recording was realized with the software "Motion Studio" (Integrated Design Tools, Inc., USA). The images were analyzed with a drop shape plugin in ImageJ (Stalder et al., 2010).

Contact angles at different temperatures were measured with the dataphysics OCA 20 instrument at the P03 beamline at DESY, Hamburg, Germany. Within the device, there are two cameras, one which captures an image of the droplet as a side view and another one that allows the droplet and sample to be seen from a top angle. A temperature controlled water bath is used to maintain the temperature of the sample holder (temperature measurement, sensor 1) and there is a climatic chamber which is placed over the sample plate to control the environmental temperature (sensor 2) around the sample during measurement. The measurement was done at two different temperatures at approximately 15 and 20 °C as average values between sensor 1 and 2. The temperature difference between both sensors were up to 5 °C due to insufficient insulation of the climate chamber. Calcium silicate was used as a desiccant inside the chamber to avoid water droplet formation due to condensation during the experiment. The contact angle changes were recorded for 18 minutes for every sample.

Liquid surface tension was measured with the Krüss (Hamburg, Germany) tensiometer K100.

3.2.3   X-ray tomography

The synchrotron based tomography experiment was conducted as described in Reinke et al. (2015c), Lügger et al. (2016). The three dimensional images of the interior structure were taken at the imaging beamline P05 operated by the Helmholtz-Zentrum Geestacht (HZG), at the Petra III storage ring at Deutsches Elektronen-Synchrotron (DESY) in Hamburg (Greving et al., 2014, Haibel et al., 2010). The principle of X-ray absorption tomography is a difference in electron density of the materials. However, chocolate components being organic materials have very similar electron densities and thus visualization is limited by density contrast. Therefore, particles in conventional chocolate

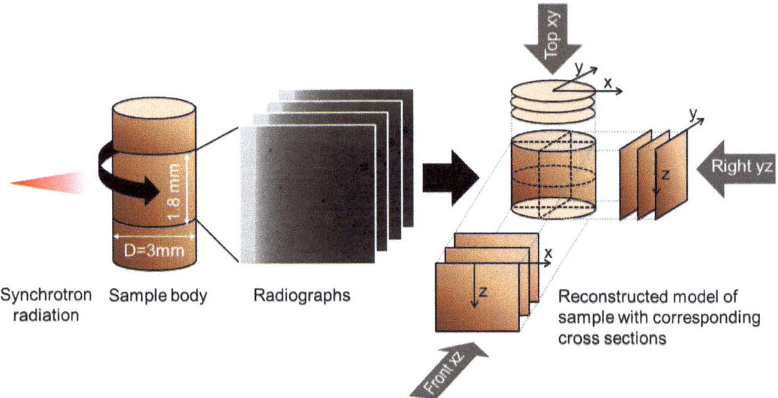

FIGURE 3.7: Experimental set-up of tomography experiment.

cannot be distinguished from the matrix phase in images taken with standard laboratory equipment within reasonable measurement time. Considering that, synchrotron radiation improves the density contrast in comparison to standard laboratory tomography devices. Absorption contrast tomography was conducted with an exposure time of 350 ms and an energy of E=11 keV (E=13 keV and an exposure time of 2000 ms were applied for a selected dark chocolate sample for better spatial resolution).

The size of the CCD (charge-coupled device) of the detector was 3056 x 3056 pixel and the optical magnification was 9.7 (for the selected dark chocolate sample magnification was set to 20.4) resulting in a field of view of 3.78 x 3.78 mm (and for the selected dark chocolate sample 1.8 x 1.8 mm). The sample was then rotated by 180° (or 360° for the selected dark chocolate samples[7]) in angular steps of $\sim 0.2°$. The effective pixelsize of the scan raw data was 1.24 µm (in case of the selected dark chocolate 588 nm). In order to achieve higher contrast, the image was reconstructed with double (2x) binning in addition to the unbinned reconstruction (Beckmann et al., 2008, Thurner et al., 2004) (Figure 3.7). Double binning gives a 2.47 µm (and 1.175 µm for the selected chocolate, respectively) resolution. 4x binning of the selected dark chocolate sample resulted in a resolution of 2.352 µm. Besides investigation of different cross sections in top xy, front xz and side view yz (Figure 3.7), the tomograms were further analyzed with the 3D and the volume viewer plug-ins of the software package Fiji (Schindelin et al., 2012, Schmid et al., 2010).

In addition, samples (with a diameter of about 5 mm and an artificial capillary with a diameter of about 350 µm at time of sample preparation) were imaged with the µCT 35 from SCANCO MEDICAL AG at the Institute of biomechanics (Hamburg University

---

[7]With a 360° scan and a laterally shifted rotation axis a sample size of 2 times the field of view can be visualized, resulting in a maximum sample size of approximately 3.6 mm.

of Technology, Germany). The radiation was 70 kV and 114 µA and exposure time was 459 ms. The total imaged height was 0.8 mm. These samples were imaged directly after preparation and after storage for 9 weeks at room temperature as well as storage at 32 °C for 3 weeks with subsequent storage at room temperature for 6 more weeks. The diameter of the artificial capillary, which has been molded into the center of the sample, was determined from the tomographic images. The diameter was taken at two positions within the scanned 3D model.

3.2.4 Simulation of stress distribution during solidification

The procedure of simulation of stress formation during solidification of chocolate was already published in Reinke et al. (2015c). The interior chocolate microstructure was modeled as a two phase system with rigid particles surrounded by a soft matrix. The distribution of particles within the matrix is based on the tomographic images of the real chocolate system. Four different systems have been modeled. A total area of 540 x 930 µm^2 was simulated in all four models. The first model is directly based on a tomographic image (Figure 5.12) with a disproportional large particle. The average Feret diameter of all particles in this 2D model is 25 µm. The area fraction of particles in the model was set to a total of 65 % including the large particle and an average area fraction of 60 % excluding the large particle. Additionally, three chocolate models were simulated with a smaller particle size distribution. Thereby, an average Feret diameter of 23 ± 2 µm and an area fraction of particles of 58 ± 1 % was modeled. The three models differ in particle shape: sharp, round and both sharp and round particles were simulated.

Meshing was performed with the software package OOF2 (Reid et al., 2009). The finite element simulation was done with the commercial software Abaqus Version 6.12 from Simulia (Dassault Systèmes, Vélizy-Villacoublay, France). Contact of two particles (which was observed in the tomographic image) was avoided in the simulation set-up, because it would be simulated as a single large particle in the simulation. Thus, all particles were surrounded by the soft lipid matrix. The properties of the rigid particles were derived from the properties of crystalline sucrose with an elasticity modulus of $E_{particle} = 35 \ GPa$ (Ramos and Bahr, 2007) and a Poisson's ratio[8] of $\nu_{particle} = 0.2$ (obtained from an uniaxial compression test of granular sugar in Molenda et al. (2006) which is in accordance with Ramírez et al. (2009)). The thermal expansion of sucrose is assumed to be neglectable in the temperature range from 10 °C to 32 °C. Typical thermal expansion coefficients for solid materials which do not significantly contract

---

[8]The Poisson's ratio is the fraction of transverse contraction strain to longitudinal extension of a material.

at temperatures between 0 °C and 40 °C are in the range of $10^{-4}\,1/K$ to $10^{-7}\,1/K$ (e.g. Bailey and Yates, 1970, Cverna, 2002, DIN ISO 7991, 1998, Otto and Thomas, 1963). Thermal expansion coefficients of $\alpha_{particle} = 10^{-4}\,1/K$, $\alpha_{particle} = 10^{-5}\,1/K$ and $\alpha_{particle} = 10^{-7}\,1/K$ have been modeled to investigate the influence of particle contraction within the expected range. The thermal expansion of the particles in the main studies where set to $\alpha_{particle} = 10^{-7}\,1/K$. The surrounding phase is modeled as a soft matrix, which contracts significantly due to the temperature change from 32 °C to 10 °C. Contraction is caused by thermal contraction of the material and volume changes because of crystallization and densification of crystals, going from a viscous liquid to a mainly crystalline, solid and brittle material. The contraction is dependent on the crystallization conditions and composition of fat phase. The contraction of cocoa butter and chocolate was investigated by Mehrle (2007). They measured a total contraction of 6 % for a temperature decrease from 32 °C to 10 °C for seeded cocoa butter. To model contraction of the matrix an apparent thermal expansion coefficient $\alpha_{matrix}(T)$ is introduced, which gives an average value of how much the matrix contracts per temperature change in the range from 32 °C to 10 °C. Values were averaged for steps of 5 °C. The apparent thermal expansion coefficient of the matrix $\alpha_{matrix}(T)$ is based on the data measured by Mehrle (2007) and listed in Table 3.1. Mehrle (2007) measured the contraction of a cocoa butter crystal suspension during solidification from 32 °C to 0 °C. With regard to material properties, cocoa butter changes from a brittle solid at 0 °C to a

TABLE 3.1: Solid fat content SFC (Torbica et al., 2006), estimated Young's modulus E_{matrix} and apparent thermal expansion coefficient α_{matrix} for the simulation of cocoa butter, based on data from Maleky and Marangoni (2011b) and Mehrle (2007)

Temperature, °C	Solid fat content SFC	E_{matrix}, MPa	α_{matrix}, 1/K
10	0.89	177	1.00E-03
15	0.85	170	0.0014
20	0.82	167	0.0026
25	0.75	157	0.006
30	0.51	121	0.002

viscous liquid at 32 °C (Engmann and Mackley, 2006). Thus, the matrix phase changes from a plastic to an elastic material with temperature decrease. Young's modulus is used to relate the stress to the deformation of an elastic material or of a plastic material until their elastic limit, respectively (Marangoni, 2000). Maleky and Marangoni (2011b) measured Young's modulus of cocoa butter and proposed a general equation to relate Young's modulus E to the solid fat content $SFC(T)$ and the fat nanostructure, which is represented by the fractal dimension D and a constant λ. The equation is based on the work of Marangoni (2000), Marangoni and Rogers (2003), Narine and Marangoni (1999b,c) (Equation 2.3). Equation 2.3 applies to fat networks with a solid fat content between 60 wt% and 100 wt% and small deformations. The parameters of Equation 2.3

are according to Maleky and Marangoni (2011b) $\lambda = 143.7$ MPa and $D = 1.41$ for cocoa butter with a nanostructure with larger fat nanoplatelets (nanoplatelets with a length of approximately 2000 nm). Smaller nanoplatelets with a size of approximately 150 nm lead to values of $\lambda = 237.2$ MPa and $D = 1.64$ (Maleky and Marangoni, 2011b). Average values of $\lambda = 190.5$ MPa and the fractal dimension $D = 1.5$ were taken for this simulation. The solid fat content is dependent on the temperature, cocoa butter origin and processing condition (El-Mallah and Megahed, 1998, Torbica et al., 2006)[9]. Torbica et al. (2006) measured the solid fat content of tempered cocoa butter with two different methods (the official IUPAC 2.150 method and Karlshamns' method, see also Figure 2.5). Average values from these two methods have been taken for solid fat content. Table 3.1 summarizes the data for Young's modulus of the matrix $E_{matrix}(T)$, which is dependent on the solid fat content and thus temperature, and the apparent thermal expansion coefficient used for the simulation. Young's moduli and apparent thermal expansion coefficients for intermediate temperatures were interpolated linearly. The Poisson's ratio of the matrix phase was set to $\nu_{matrix} = 0.15$, because measured data for Young's and shear modulus from Maleky and Marangoni (2011b) suggest that Poisson's ratio of cocoa butter is rather small.

The dark chocolate mass in the experiment, on which the simulations are based, is cooled down from a liquid at about 50 °C to 10 °C for chocolate solidification. It is assumed that stresses at temperatures higher than 25 °C are immediately relaxed by displacement of the liquid fat phase. Cocoa butter starts to significantly change from a more viscous liquid to form a solid network, which can bear stresses, as solid fat content and Young's modulus increase with decreasing temperature (Table 3.1). Therefore, stress simulation was conducted in three steps from 25 °C until 10 °C. A homogeneous solidification through the structure is assumed.

3.2.5 Visual observation of macroscopic migration

Migration was observed on a macroscopic level based on visual detection of the height change of a stained layer within a sample. The experimental set-up is based on Maleky and Marangoni (2011a), Marty et al. (2009) (system B) and was further developed to system C and D. The height of the colored front can be easily detected and measured with a ruler for the porous glass cylinder (system A) and in system B due to light color of unstained cocoa butter parts and the transparent color of the porous glass. The initial lower and upper edge of the stained layer in systems C and D are marked at the outside of the transparent cylinder, in which the samples were molded and stored, directly after sample preparation. From the initial height, height change was measured with a ruler.

---

[9]Further information on the solid fat content is presented in Section 2.2.2.4 on page 16.

The migration height of the filling into the porous glass was normalized with the contact angle to better compare migration within the porous glass to that through cocoa butter and chocolate. Contact angles were measured with the sessile drop technique (Section 3.2.2 on page 47). Contact angle of a quasi static sessile droplet of sunflower oil on the surface of the porous glass cylinder is $19 \pm 3°$. Contact angle of sunflower oil on cocoa butter is about $55°$ and thus higher than that of an oil droplet on glass. Migration height of the glass cylinder was thus normalized to the contact angle of oil on cocoa butter (Equation 3.6).

$$h_{CB} = h_{exp} \cdot cos \left(\frac{\theta_{CB}}{\theta_{glass}} \right) \tag{3.6}$$

3.2.6 Simulation of migration

Migration heights were calculated based on the calculations presented in Grüner et al. (2016) (Equation 2.11) with MATLAB, Release 2013b, The MathWorks, Inc., Natick, Massachusetts, United States. The properties of the matrix phase and the migrant can be combined into two constants a and b, so that migration height is proportional to the square root of time (Equation 3.7).

$$h(t) = \sqrt{2a \cdot t + b} \tag{3.7}$$

Equation 3.8 describes the rate of migration a with the permeability K (Equation 2.12) and pressure gradient Δp (Equation 2.8).

$$a = \frac{K \cdot \Delta p}{\eta \cdot \phi_i} \tag{3.8}$$

Constant b (Equation 3.9) gives information on the initial conditions with the height h_0 at the time t_0.

$$b = \frac{h_0^2 - 2a \cdot t_0}{2} \tag{3.9}$$

The parameters of the moving liquid are approximated with macroscopic properties of liquid sunflower oil. The parameters in the model are the surface tension $\gamma = 35$ mN/m, viscosity $\eta = 60 \cdot 10^{-3}$ Ns/m^2 and density $\rho = 925$ kg/m^3.

First, migration heights of lipids into porous glass are calculated and compared to experimental data in order to extract the hydrodynamic radius r_h (Figure 3.8). The hydrodynamic radius is the radius at which the flow pattern develops and might be larger (in case of slippage) or smaller (in case of sticking of immobile molecules at the pore wall) than the channel radius r_0 (Grüner et al., 2016). The difference in the two radii can be expressed by the ratio f_h (Equation 3.10).

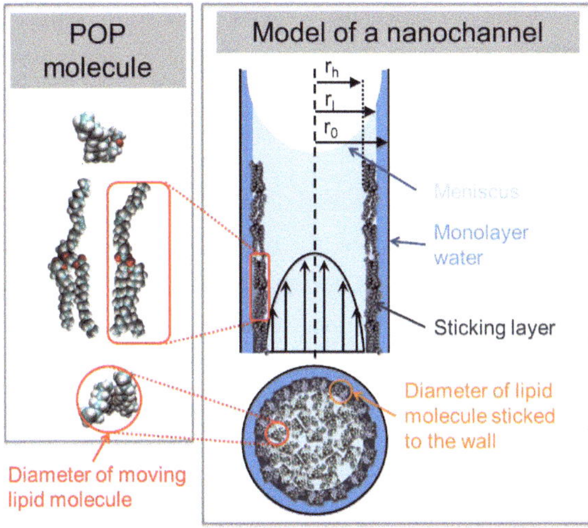

FIGURE 3.8: Schematic overview over model to simulate lipid migration in a nanochannel based on Grüner et al. (2016).

$$f_h = \frac{r_h}{r_0} \tag{3.10}$$

The porous glass has a pore size distribution with $r_0 = 3.4$ nm as modal value, a tortuosity of $\tau = 3.6$ and volume porosity of $\phi_0 = 0.3$, which is reduced due to water absorption (Grüner et al., 2016, Lin et al., 1992). Grüner et al. (2016) measured a reduction in ϕ_i with increasing relative humidity. The initial volume porosity is reduced to $\phi_i = 0.245$ at a relative humidity of 50 %. The diameter at which the meniscus forms r_l is reduced due to sticking of a monolayer water at the wall to $r_l = r_h - 0.25$ nm (Grüner et al., 2016). The tortuosity of the porous glass is 3.6 (Grüner et al., 2016). The equations were solved in MATLAB, Release 2013b, The MathWorks, Inc., Natick, Massachusetts, United States. The matlab code for simulation of capillary rise through confined spaces of the porous glass with variations of f_h is attached in the Appendix on page 135.

The number of molecules n_{mol}, which flow through the channel, is estimated based on the ratio between the cross sectional area of flow with the radius r_h and that of a molecule with a radius of r_{mol} (Figure 3.8 and Equation 3.11). It is assumed that lipid molecules arrange into flow direction (similar to that of particles in microchannels as observed by Trebbin et al. (2013) and suggested by Grüner et al. (2016) for molecules).

$$n_{mol} = \frac{A_{cap,mol}}{A_{mol}} = \frac{A_{cap} \cdot (1 - \phi)}{A_{mol}} = \frac{r_h^2 \cdot (1 - \phi)}{r_{mol}^2} \tag{3.11}$$

The packing density of molecules $(1 - \phi)$ is assumed to be 0.74 (highest packing density) within the estimation. The dimensions of a triglyceride molecule were estimated based on published data of crystalline cocoa butter as well as a molecular simulation of a single POP molecule in vacuum.

Second, the impact of contact angle, radius, tortuosity and viscosity on capillary rise of oil were analyzed. Therefore, the tilt angle of the capillaries are kept at 45° to account for the average between horizontal and vertical capillaries. Migration started at $t_0 = 0.98$ days as initial conditions, which is the time at which linear increase of migrated height with the square root of time starts (Figure 6.4). Contact angle, capillary radius and viscosity are varied in order to analyze their influence on migration height (Figure 6.27). Viscosity was set to 60 mPa which is in the range of the macroscopic viscosity of food oils such as sunflower and coconut oil (Bürkle, 2011). Capillary radius was set to 3.4 nm based on Grüner et al. (2009). A macroscopic contact angle of $\theta = 19.3 \pm 3.4°$ was measured by sessile drop technique based on a sunflower oil droplet on top of the outer surface of the porous glass. The dynamic contact angle on a nanoscopic level within the capillaries cannot be easily determined (e.g. Reinke et al., 2015a) and thus the macroscopic contact angle is used as approximation.

Third, experimentally measured migration in tempered and untempered cocoa butter were fitted to Equation 2.11 (on page 29) in order to estimate which values of a (Equation 3.8) as well as c (Equation 3.9) describe the migration in tempered and untempered cocoa butter. In addition to an overall fit, fits were conducted in different ranges of time in order to describe the stepwise height increase.

An effective diffusion coefficient D_{eff} can be estimated from the slope of migration height per square root of time based on Equation 2.14 (page 31). The sample geometry is a cylinder with the volume V, a height h_{sample} and a cross sectional area of A. This leads to Equation 3.12.

$$D_{eff} = \left(\frac{h(t)}{\sqrt{t}} \cdot \frac{h_{sample}}{h_{sat}} \right)^2 \tag{3.12}$$

Under the assumption that the saturation height h_{sat} is the sample height h_{sample}, an effective diffusion coefficient can be determined by squaring the slope $\frac{h(t)}{\sqrt{t}}$.

Combining Equation 3.12 and Equation 2.14 (page 31) leads to Equation 3.13 by which a diffusion coefficient D_{sim} from the simulated migration height can be extracted under the assumption that the constant b can be neglected.

$$D_{sim} = a_{sim} \cdot 2 \cdot \left(\frac{V}{A \cdot h_{sat}} \right)^2 = a_{sim} \cdot 2 \cdot \left(\frac{h_{sample}}{h_{sat}} \right)^2 \tag{3.13}$$

Under the assumption that $h_{sat} = h_{sample}$ it would give that $D_{sim} = 2 \cdot a_{sim}$.

A theoretical capillary radius r_{sim} can be extracted from the simulated data of a_{sim} by Equation 3.14 with the initial radius, which was set to $r_0 = 500$ nm and $r_0 = 10$ nm at the beginning of the simulation[10], and an initial value for the constant a (Equation 3.8) which is a_0.

$$r_{sim} = r_0 \cdot \sqrt{\frac{a_{sim}}{a_0}} \qquad (3.14)$$

3.3 Additional analytic methods

Differential scanning calorimetry for melting temperature determination and scanning electron microscopy was applied in addition to the main experiments.

3.3.1 Differential scanning calorimetry

Differential scanning calorimetry (DSC) measurements have been conducted with a NETZSCH DSC 204 F1 Phoenix device, Netzsch, Selb, Germany. The samples were first cooled down to 15 °C and subsequently heated up from 15 °C to 50 °C at 1 K/min. The sample mass was 13.5 ± 1.0 mg. The enthalpy of melting δH was determined by integration of the melting peak from 22.5 °C to 35 °C and a linear baseline was applied. The intergrations and calculations were performed with Origin (OriginLab, Northampton, MA).

3.3.2 Scanning electron microscopy

Scanning electron microscopy was used to image the powder samples using a Zeiss Supra 55 VP FEG-REM, Carl Zeiss, Oberkochen, Germany. The aperture size was 10 µm and the secondary electrons were detected. The voltage was adjusted from 1 kV to 3 kV according to optimum image quality.

---

[10]Different starting values of the initial radius r_0 were tested and the chosen value allowed for fitting with MATLAB, Release 2013b, The MathWorks, Inc., Natick, Massachusetts, United States. However, the fitting process should be optimized by e.g. evaluation of other solvers in order to better represent the experimental data. Nevertheless, the calculations allowed for a first approximation of experimentally measured migration heights with the model presented in equation 3.7.

4

Tracking oil migration into chocolate powders

The aim of the small angle X-ray scattering study was to identify preferred pathways and track structural changes during migration of oil into chocolate components. The results are the basis to evaluate potential pathways through chocolate and lipid migration mechanisms. The fat phase in chocolate is partly liquid and partly solid at room temperature. Thus, migration might take place through either the network of particles embedded in the matrix phase or through the fat phase or at the interface of both phases. In addition to the particle and fat phase, structural defects filled with air such as pores and crevices are possible migration pathways. The driving force is linked to the pathway[1].

Observation of oil migration in-situ without sample destruction provides further details towards a better understanding of migration in multicomponent food products. Therefore, the pathway of oil migration in multi-component food materials containing cocoa butter as a fat phase was investigated with synchrotron small angle X-ray scattering (SAXS). This technique provides information about pores and crystal structure at the same time. With the used set-up structures in a range of about 3.5 nm to 125 nm are detected. The samples were analyzed in its natural state (denoted as dry powders) and after oil migration to identify structural changes induced by oil migration. The SAXS pattern provides information on the material structure in the form of Bragg peaks which result from diffraction at the highly ordered crystalline lattices and regular structures such as particles or pores in the nm range. The use of powders instead of solid chocolate bars enables faster migration within beamtime which is up to a day rather than weeks or months, which are typically migration rates in chocolate (Maleky and Marangoni,

---

[1]See section 2.4 on page 26 for a summary of different transport mechanisms and constraint of space.

2011a, Walter and Cornillon, 2002). Furthermore, single powders with and without cocoa butter instead of mixtures usually present in real food products are analyzed to reduce complexity.

Parts of this study were published in Reinke et al. (2015b).

4.1 Data presentation from small angle X-ray scattering of chocolate powders before and after oil migration

First, the structure of five chocolate components, namely skim milk powder with and without cocoa butter, pure cocoa butter, sucrose with cocoa butter and cocoa powder, which is cocoa solids with cocoa butter, were analyzed. The sucrose and skim milk powder with cocoa butter were in form of refiner flakes. The five dry powders (including the two refiner flakes) before migration of oil show three q-regions corresponding to ordered structures in the SAXS patterns (Figure 4.1). The peaks at $q = 0.97\,\text{nm}^{-1}$ and $q = 1.39\,\text{nm}^{-1}$ are the characteristic peaks of cocoa butter forms V and IV, respectively. The corresponding d-spacings are 6.50 nm and 4.50 nm. Wille and Lutton (1966) and Chapman et al. (1971) reported a d-spacing for cocoa butter in form IV ranging from 4.50 nm to 4.90 nm, for form V from 6.41 nm to 6.60 nm and for cocoa butter in form VI from 6.30 nm to 6.38 nm. Only the cocoa butter IV and V polymorphic forms were observed in the samples with cocoa butter. However, the peak for form V is close to the VI peak and thus might overlap with the V peak. However, the DSC (differential scanning calorimetry) measurement (Figure 4.2) clarifies that form V is predominant in the samples. The shoulder at a q-value of around $0.4\,\text{nm}^{-1}$ corresponds to a d-spacing of 15.84 nm (Figure 4.1) and is well-known for milk products. It is characteristic of the supramolecular structure of casein, which is present in skim milk powder (Gebhardt et al., 2010b, 2011). Gebhardt et al. (2010b) measured casein films and found shoulders at $q = 0.03\,\text{nm}^{-1}$ (attributed to the size of casein micelles), at $q = 0.1\,\text{nm}^{-1}$ (arising from micellar substructures, so-called mini micelles), and at $q = 1\,\text{nm}^{-1}$ (attributed to the colloidal calcium phosphate) in their study. Shukla et al. (2009) and Holt et al. (2003) observed a shoulder near $q = 0.35\,\text{nm}^{-1}$. They attributed the shoulder to calcium phosphate nanoclusters. However, the exact internal structure of casein micelles is still not fully understood (Gebhardt et al., 2010a,b, 2011). In our case, the results do not show pronounced peaks for skim milk powder besides that at $q = 0.4\,\text{nm}^{-1}$ (Figure 4.1).

The differential scanning calorimetry (DSC) measurements of selected samples show a prominent melting peak with an onset temperature of 22.9 °C and a maximum at 32.4 °C for cocoa powder (Figure 4.2). The onset temperature for the refiner flakes with

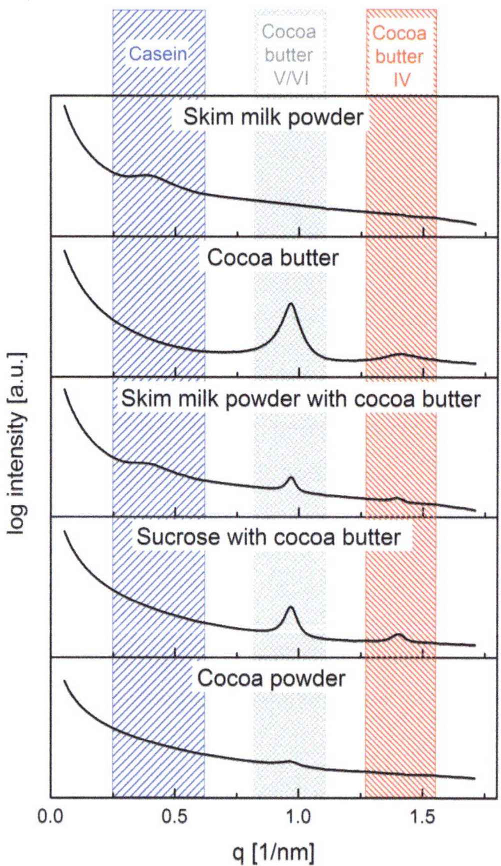

FIGURE 4.1: Scattering curves of dry powders, sum of vertical scan (images 1-61).

cocoa butter and sucrose as well as the ones with skim milk powder and cocoa butter is 22.6 °C. Temperatures at maximum heat flow are 32.9 °C for the sucrose refiner flakes and 33.1 °C for refiner flakes with skim milk powder. The melting peak of cocoa butter starts at 25.0 °C and has its maximum at 32.7 °C. The melting temperature range for cocoa butter form IV is at about 25.6 °C (Chapman et al., 1971) to 27.5 °C (Lonchampt and Hartel, 2004) whereas the form V melting temperature varies from 30.8 °C (Chapman et al., 1971) to 33.9 °C (Lonchampt and Hartel, 2004). Form VI melts at about 32.3 °C (Chapman et al., 1971) to 36.3 °C (Lonchampt and Hartel, 2004). The cocoa butter form IV crystals cannot be clearly identified in the DSC measurements of the cocoa butter-powder mixtures as they are probably merged with the form V melting peak. Cocoa butter has a broad range of melting temperatures, which is dependent on the exact composition and thermal history (Chapman et al., 1971, Lonchampt and Hartel, 2004, Metin and Hartel, 2005). The melting enthalpy (Table 4.1) of pure cocoa

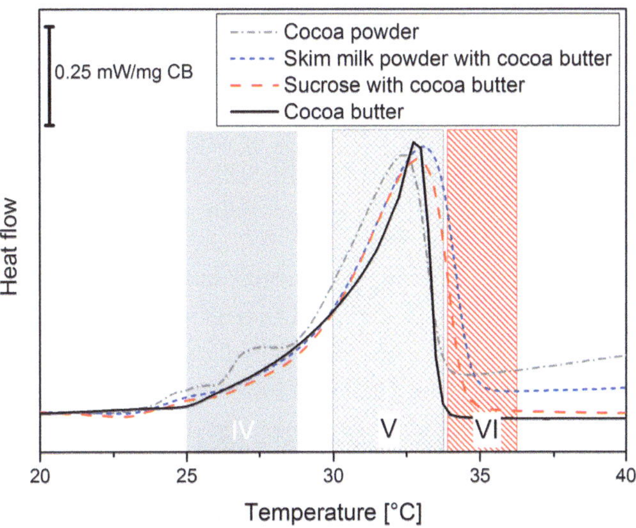

FIGURE 4.2: Differential scanning calorimetric measurement of selected samples. The melting temperature ranges for different cocoa butter polymorphic crystal forms are indicated in the figure.

butter is 128.7 J/$g_{cocoabutter}$ and for cocoa powder it is 127.2 J/$g_{cocoabutter}$, respectively. This is in accordance to Afoakwa et al. (2008b). They measured melting enthalpies of tempered chocolate samples with varying particle content and particle size between 110 J/g_{fat} and 130 J/g_{fat}. The melting enthalpies of cocoa butter and cocoa powder are lower than those of the refiner flakes which are 152.3 J/$g_{cocoabutter}$ for the skim milk powder with cocoa butter and 153.2 J/$g_{cocoabutter}$ for sucrose with cocoa butter. Thus, the crystal content and consequently solid fat content of cocoa butter is higher for the refiner flakes. The cocoa butter and cocoa powder are in its natural state, whereas the refiner flakes were melted and solidified during production process. The sucrose and skim milk powder particles might thereby have acted as heterogeneous nucleation agents for crystallization of the cocoa butter matrix, leading to higher crystal content. However, solidification process was uncontrolled and crystals might also have been rearranged during storage time due to insufficient tempering. This is supported by Figure 4.2 which shows that the melting peaks of the refiner flakes are shifted slightly to higher temperatures compared to cocoa butter and cocoa powder and part of the melting peak is in the area of the form VI melting range. It could thus be that the refiner flakes contain more form VI crystals at time of measurement than cocoa powder and butter and following the melting enthalpies are higher. Afoakwa et al. (2009b) measured melting enthalpies of under-, over- and well tempered chocolate with a fat content of

TABLE 4.1: Melting temperatures and melting enthalpies of cocoa butter in powder samples extracted from DSC.

Material	Onset temperature [°C]	Peak center [°C]	Enthalpy of melting [J/g CB]
Cocoa butter (CB)	25	32.7	128.7
Cocoa powder	22.9	32.4	127.2
Skimmed milk powder with CB	22.6	33.1	152.3
Sucrose with CB	22.6	32.9	153.2

35 wt% after 14 days of storage and melting enthalpies of well tempered samples were in the range of 105 J/g_{fat} to 108 J/g_{fat}, whereas under tempered samples showed higher values of 125 J/g_{fat} to 127 J/g_{fat} accompanied with a slight shift of melting peak towards higher temperatures (about 2 °C higher peak temperature). It can be seen that the most prominent peak in the cocoa butter-powder mixtures is the cocoa butter form V peak at $q = 0.97\,\text{nm}^{-1}$ (Figure 4.1) indicating predominance over form IV and VI. Forms IV and V are both stable at room temperature, but form VI is thermodynamically more favorable. Therefore, form V and IV transform into form VI over time (Metin and Hartel, 2005). The form IV peak for cocoa powder is not visible in the SAXS measurements even though a small but identifiable peak in the form IV polymorph can be observed in the DSC results for cocoa powder (Figure 4.2).

Apart from the identification of cocoa butter crystal structure, the SAXS data were also used to gain a deeper insight into the distribution of different polymorphic forms (Figure 4.3). Cocoa butter form V is predominant in every measured point, whereas the crystalline cocoa butter polymorphic form IV is heterogeneously distributed in the sample. Form IV peak at $q = 1.39\,\text{nm}^{-1}$ is not present at every spatial point (Figure 4.3). Thus, the SAXS patterns clearly demonstrate that the cocoa butter sample posses crystal grains in different polymorphic forms. The spatial distance between the center of the X-ray beam of two adjacent scattering points is 100 µm with a beamsize of 22 µm in the vertical direction. Thus, two adjacent points with cocoa butter form IV identify grain sizes of up to 122 µm. In contrast, the sucrose with cocoa butter sample showed in every measured spatial point equally distributed polymorphic forms IV and V with smaller peak width for the IV peak. The skim milk powder with cocoa butter showed identifiable cocoa butter form IV peaks in around 75 % of the investigated spatial points. The DSC measurements (Figure 4.2) show that the melting peak of the refiner flakes is broader than the one of pure cocoa butter. This suggests that the cocoa butter in the refiner flakes crystallized into polymorphic forms IV to VI, whereas pure cocoa butter mainly contains form V and some crystal grains in form IV (Figure 4.1 to 4.3). Additionally, particles might enhance more homogeneous crystallization which explains the homogeneous distribution of form IV peaks in case of the refiner flakes (cocoa butter

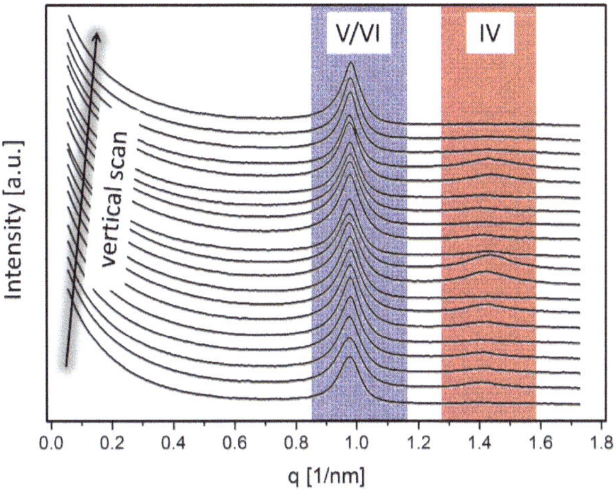

FIGURE 4.3: 1D scattering patterns of cocoa butter before oil migration (dry powder) from positions 1 to 20 in the vertical scan. The q-ranges of form IV and V/VI are indicated in the figure.

with sucrose and skim milk powder particles). In contrast, the pure cocoa butter sample has some but large crystal grains in form IV which are more heterogeneously distributed.

To study oil migration, the complete set of 61 images of every vertical scan was summed up in order to get results which are representative of the heterogeneous sample at each defined oil migration time. Thus, the new set of images with 1 image per time point for all samples represents the cumulated intensity over the vertical scan. The significant scattering for all powders before oil addition (Figure 4.4) reveals the presence of regular structures of the dry powders and refiner flakes in the measurement range (3.5 nm to 125 nm) (Porod, 1982). In case of the examined powders this is most likely a result of pores which are present in that range (Figure 4.5 and Figure 4.4) before the addition of oil. The scanning electron images (Figure 4.5) show that especially cocoa powder and the refiner flakes with sucrose and cocoa butter have a structure with pores and particulate objects of sizes in the range of 100 nm which is in the measured SAXS regime[2] . Cocoa powder is a highly porous network of cocoa solid material coated with cocoa butter. The surface of a cocoa powder particle has many pores in the SAXS range used in this study (which is from 3.5 to 125 nm) and the entire structure is highly fractal. The porous structure of the powders is also visible in the scanning electron microscopy images (Figure 4.5). On the other hand, the structure of skim milk powder is less heterogeneous and smoother. There are pores in the measured SAXS range as well as it can be seen in Figure 4.5, however less compared to cocoa powder. This explains

---

[2]It was not possible to get sharp images with scanning electron microscopy of structure at even higher magnifications.

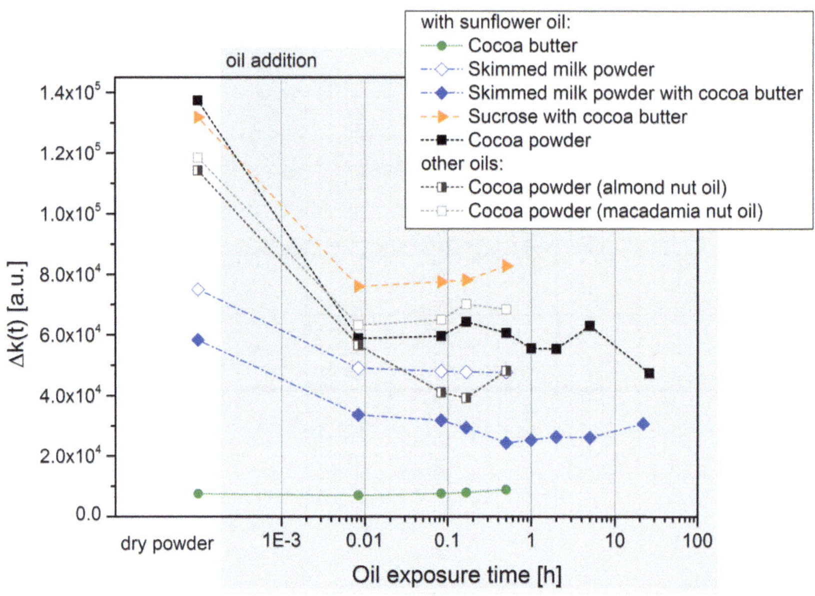

FIGURE 4.4: Difference of total scattering power of different powders in dependence of migration time.

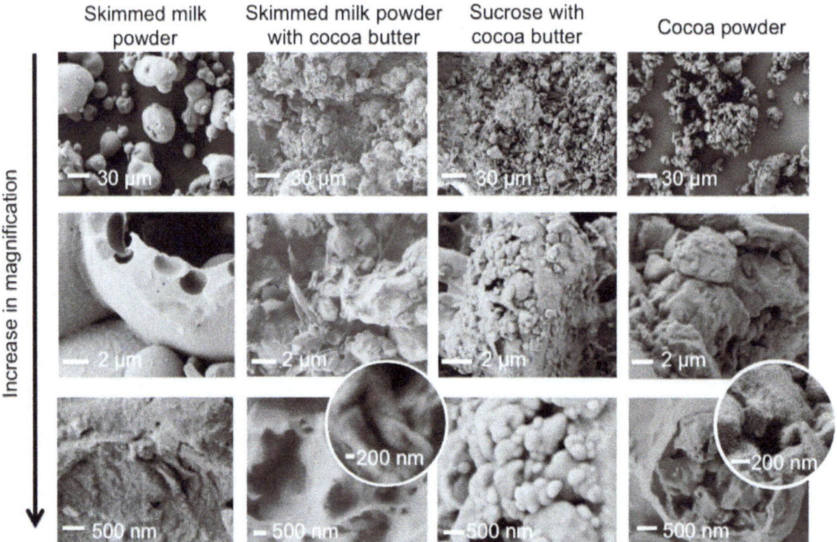

FIGURE 4.5: SEM images of sucrose refined with cocoa butter, cocoa powder, skim milk powder, and skim milk powder refined with cocoa butter.

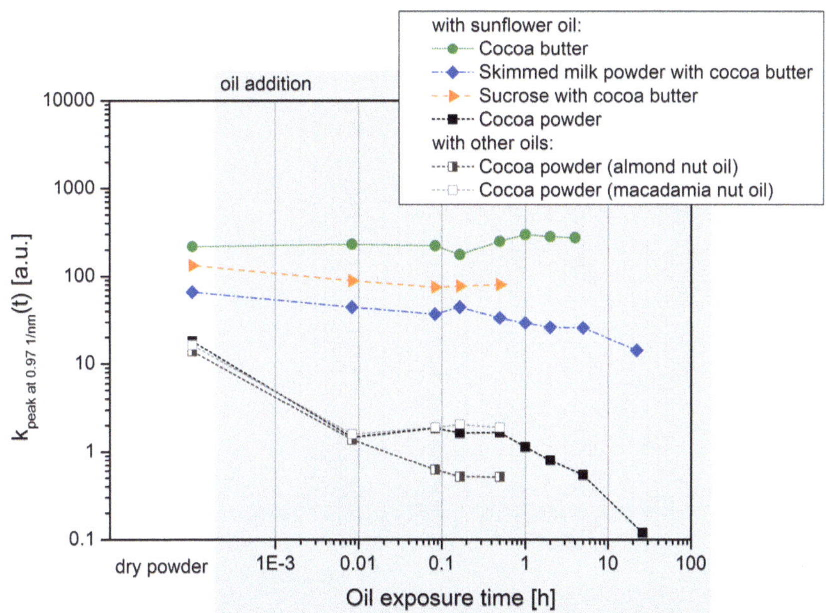

FIGURE 4.6: Peak intensity of cocoa butter peak (form V/VI) at different oil exposure

the lower Δk. In the case of skim milk powder refiner flakes with cocoa butter it can be seen that Δk decreases, which is probably due to blocking of the pores because of the cocoa butter coating. This can also be seen in Figure 4.5. Pure cocoa butter shows lower Δk values probably due to its less porous structure as compared to the other samples. Due to addition of oil, Δk is decreasing for every powder sample but remains almost constant for the cocoa butter sample (Figure 4.4). The most significant decrease in Δk takes place in the first minutes and further decrease slows down with time.

Besides the preceding results, the integral intensity of the crystalline peak associated to cocoa butter form V ($q = 0.97\,\mathrm{nm}^{-1}$) decreases gradually with time, which means that the crystalline structure vanishes probably due to dissolution (Figure 4.6). Unfortunately, due to the weak signal of the peak associated with form IV, the expected changes in the intensity of this peak could not be followed.

The analysis of the single scattering patterns from each spatial point[3] (Figure 4.7) for all samples revealed that structural changes occurred in the same magnitude over all 61 points of the vertical scan. Intensities are however distributed over the scanned area which is due to heterogeneity of the sample[4].

---

[3] All imaged spatial points were averaged in Figures 4.4 and 4.6. 20 of the single patterns are shown for example in Figure 4.3.

[4] This was analyzed for all samples and are exemplary shown for cocoa butter in Figure 4.7.

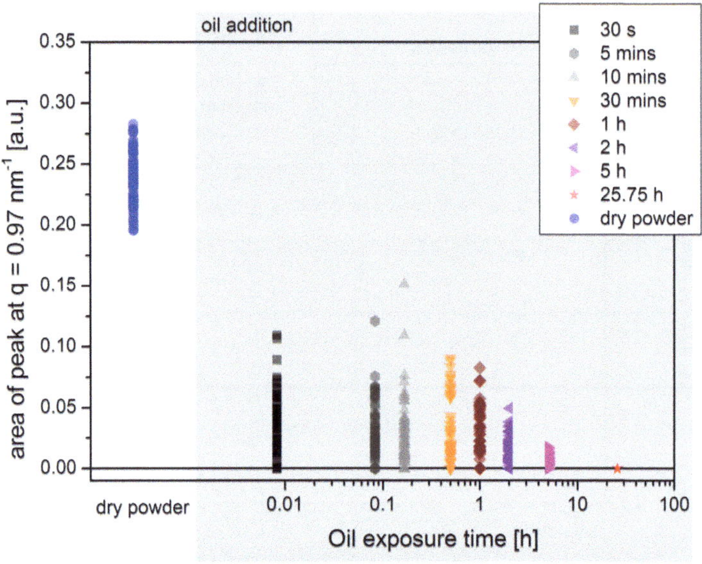

FIGURE 4.7: Area of cocoa butter polymorphic form V peak of each of the 61 single spatial 1 D scattering patterns of cocoa powder.

These results reveal that the oil migrates into the pores within the first minutes. A decrease in Δk can be explained considering a change in X-ray scattering contrast due to oil migration, since the contrast is different between the solid material and air as compared to the difference between chocolate powders and oil. According to Equation 3.2 (page 46) introduced by Porod (1982), a change in k and consequently Δk (Equation 3.5 on page 46) occurs whenever the scattering contrast changes, assuming a constant volume of material which scatters. The Δk value of cocoa butter does not change significantly, which is due to the fact that presumably there are no, or not many, porous structures within the detection range (3.5 nm to 125 nm) and therefore no significant change in scattering contrast takes place indicating that oil migration occurs predominantly through the porous structures in the first few minutes.

The decrease in the cocoa butter form V peak implies that oil migration is provoking, in the long run, a loss of the crystalline structure. The oil migration into the cocoa butter fat phase might lead to a mixture of these two lipid materials, hence the solid fat content decreases. Zeng et al. (2002) measured a solid fat decrease from 80 % to 20 % at 24 °C in a cocoa butter with oil mixture when the concentration of almond oil is increased to 50 % in a mixture of cocoa butter and almond oil (Lonchampt and Hartel, 2004, Zeng et al., 2002). Furthermore, parts of the cocoa butter might dissolve in the oil (Lee et al., 2010, Rothkopf and Danzl, 2015, Smith et al., 2007). The peak of pure cocoa butter does not significantly decrease within 5 hours, which might be due to little surface area available for fat dissolution by the migrated oil in case of pure cocoa butter. In contrast, the cocoa powder and refiner flakes samples present higher porosity or roughness, respectively (Figure 4.4). It is not possible to distinguish between surface roughness or internal pores from the given SAXS data in Figure 4.4. It might be that more surface area is available and the dissolution process is consequently accelerated. This is probable because of different accessibility to cocoa butter within the sample. Thus, better accessible parts dissolve faster.

4.2 Wetting of cocoa butter composites with sessile oil droplet

Contact angle measurements on a macroscopic scale strengthened the findings from the previous discussed SAXS study investigating micro- and nanoscale phenomena. There is a decrease in the contact angle due to spreading on the sample (Figure 4.8) within the first minutes. The oil droplet on the surface of the chocolate model system decreases up to 0.03 µl in volume for 12.5 minutes (denoted as first phase) after droplet deposition (Figure 4.9) at room temperature ($15 \pm 5°C$) and up to 0.06 µl at higher temperatures ($20 \pm 5°C$). For samples with embedded sucrose an additional volume loss of 0.02 µl from 12.5 to 18 minutes (denoted as second phase) can be seen (Figure 4.9). The initial volume loss within the very first minutes could be due to spreading or migration into surface roughness. Surface topology can be altered when particles are embedded (Reinke et al., 2015a). However, the additional decrease in volume in the second phase (from 12.5 minutes on) at both temperatures could also be the result of migration into the bulk phase by dissolution. Migration is thereby facilitated due to addition of particles (Figure 4.9). Furthermore, the rate of migration and contact angle change is dependent on temperature. The higher the temperature, the higher the change in both volume and contact angle of the sunflower droplet (Figure 4.8 to Figure 4.9). Results for pure cocoa butter and sucrose suspended in cocoa butter are presented here. However, the findings are reproducible with addition of other particles and chocolate (Reinke et al., 2015b). The embedded particles did not detectably influence the wetting and migration of sessile oil droplets within the first 12 minutes. The contact angle starts at a slightly larger value with addition of skim milk powder as compared to the other samples. This sample then decreases slightly faster. This behavior is characteristic of rougher surfaces, which might be due to the embedded particle phase. Reinke et al. (2015a) discusses the measurement of contact angle on chocolate and chocolate model systems in detail including a study of the influence of roughness on contact angle. Furthermore, the starting contact angle

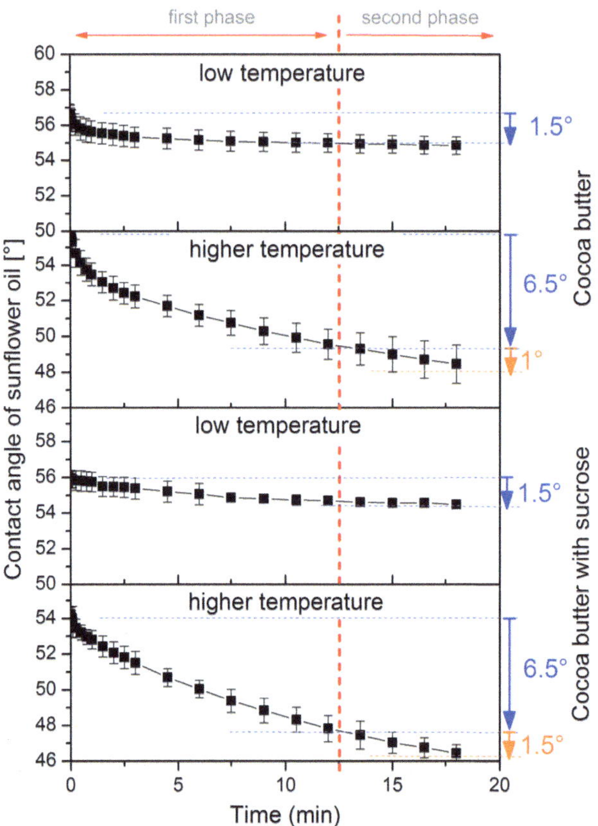

FIGURE 4.8: Contact angle of sessile sunflower oil droplet on chocolate model systems. Comparison of lower temperature ($15 \pm 5°C$) and higher temperature ($20 \pm 5°C$).

value for sucrose is slightly lower compared to the other materials, which might be due to hydrophilic nature of sugar.

Light microscopy images provided further insight into changes due to the presence of oil at the surface of chocolate model systems. The surface was imaged with a light microscope before and after 5 hours from sessile drop exposure when the droplet disappeared completely (Figure 4.10). The regular lined original shaped surface originated from molding process. The surface of both chocolate model samples significantly changed in morphology due to oil exposure. For comparison, the experiments were repeated with water. In this case, only particle swelling of cocoa powder was observed but no morphological changes of pure cocoa butter surface are found.

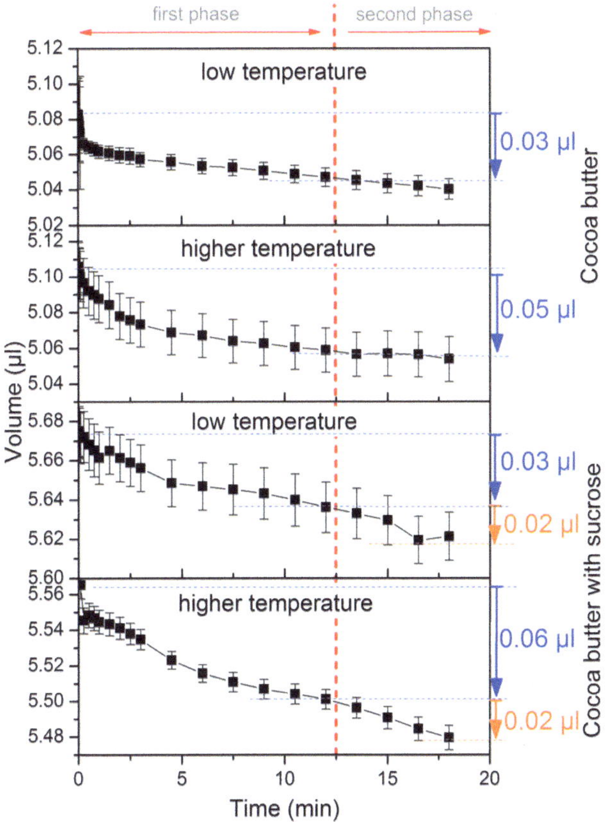

FIGURE 4.9: Volume of sessile sunflower oil droplet on chocolate model systems. Comparison of lower ($15 \pm 5°C$) and higher temperature ($20 \pm 5°C$).

To conclude, the contact angle measurements of a sessile droplet on cocoa butter and cocoa butter composites strengthened the findings from the SAXS measurements. Sunflower is wetting the material within minutes. Furthermore, it seems like there are two mechanisms overlapping with wetting which is predominantly in the short term and migration into the materials after 12.5 minutes which is dependent on accessibility and seems to be enhanced when particles are embedded. The light microscope images of the cocoa butter surfaces macroscopically reflect the oil migration through the cocoa butter in the long term.

FIGURE 4.10: Microscopy images of sample surface prior and after oil and water droplet deposition and disappearance. Cocoa butter and cocoa butter with cocoa powder are both shown.

4.3 Discussion of possible lipid migration pathways and mechanisms revealed with small angle X-ray scattering

In summary, the results suggest that oil is first migrating through pores and cracks in the solid structure probably driven by capillary pressure. Subsequently, chemical migration through the fat phase takes place inducing softening and partial dissolution of the crystalline cocoa butter. Consequently, lipid migration in chocolates probably involves both capillary rise, to be considered as the dominant mechanism in the short term, and molecular migration in the long term. The most probable migration pathway due to wetting occurs through the porous structure of the material. This process could be prevented by a reduction of porosity and a minimization of defects in the chocolate matrix, which might act as pathways. Chemical migration of lipids significantly contribute to the overall process at longer observation times. This is probably highly dependent on the availability of liquid lipids and accessibility of the cocoa butter crystals. The effect of dissolution and chemical migration could be minimized by ensuring a reduced content of non-crystallized liquid cocoa butter.

Hondoh et al. (2016) recent study on oil migration through chocolate strengthened the findings of this study. They visualized oil migration in chocolate using scanning electron microscopy-energy dispersive X-ray spectroscopy. They compared migration of silicone oil, which does not significantly dissolves cocoa butter, to that of canola oil, a vegetable oil in which cocoa butter is soluble. In addition to scanning electron microscopy they measured the weight gain over storage time. They found a higher increase in weight with silicon oil, which indicates faster migration as long as counter diffusion can be neglected. They explain fast migration with lower viscosity. However, another aspect is the possible structure change due to oil migration because of high solubility of cocoa butter in canola oil compared to silicone oil. This is strengthened with Hondoh et al. (2016)'s results from a penetration experiment which shows that the chocolate sample in vicinity to canola oil is less hard than the reference sample (pure chocolate) and chocolate with silicone oil. They also conclude from a quantitative analysis that migration and diffusion both act as independent migration mechanisms.

5

Visualization of possible pathways for lipid migration in crystalline fat suspensions

Crevices and pores in chocolate might act as pathways for convective flow driven by capillary pressure. Furthermore, in case of diffusion, the size of structure influences the type of diffusive flow. Free diffusion only occurs in ranges where the dimension of the migration pathways is significant larger than that of the migrant. When sizes are smaller, surface diffusion or diffusion through solid material are predominant. Thus, the microstructure is the basis to validate the proposed migration pathways and mechanisms. Consequently, a better understanding of the microstructure can finally lead to strategies to reduce migration and resulting deterioration of the chocolate product (Hartel, 1999). Thereby, a non-destructive visualization of the inner microstructure of chocolate is advantageous because techniques which require sample destruction might cause structural changes during its preparation. A promising technique in that regard is tomography which enables visualization of inner structures without sample destruction. As chocolate is a multicomponent material with particles embedded in a fat matrix phase, microstructure formation is complex. Therefore, structure analysis of chocolate model systems are of interest in addition to conventional chocolates. Chocolate model systems can reduce the complexity and by that facilitate the understanding of structure formation. Images of cocoa butter in its pure state and with increasing amount of suspended particles (in this case sugar) support the understanding of chocolate microstructure. Cocoa butter, which constitutes the matrix phase, determines the chocolate properties to a large extent. Thereby, crystallinity is a highly important characteristic. Thus, precrystallized tempered samples are compared to untempered samples to understand the impact of the crystallization process on the microstructure. Furthermore, the structure of fresh samples are compared to samples stored for up to a year to identify changes in the microstructure with storage time.

Parts of this study were published in Lügger et al. (2016), Reinke et al. (2015c).

5.1 Comparing the microstructure of tempered and untempered cocoa butter

Clear differences in the microstructure of tempered and untempered samples are detected with X-ray tomography imaging (Figure 5.1). The microstructure of tempered cocoa butter is rather homogeneous (Figure 5.1) without large pores. Small air bubbles or cavities throughout one of the tempered cocoa butter samples were locally found (Figure 5.1). This is probably due to insufficient air release during production. The air bubbles are very small in size (less than 10 µm). The image shows the entire section which was imaged. The sample is inside the cylindrical mold and its diameter is about 3 mm. The surface of the tempered sample is concave shaped. This is probably due to contraction during the solidification process. Tempered samples contract significantly more than untempered samples which show a quite horizontal surface. Solidification most likely starts at the outer surface and prolongs into the inner of the sample. Cocoa butter contracts during solidification. That might explain why the surface shows a minimum in the center. Additionally, it could be that the concave shape results from the meniscus of the liquid chocolate mass due to surface tension during the molding process. However, the latter would not explain why the surface of untempered samples is not concave. Tempered cocoa butter has a porous structure close to the surface layer (up to 250 microns below the surface, Figure 5.1). As opposed to tempered samples, the structure of untempered cocoa butter is heterogeneous on a micrometer level. Large cavities of about 200 to 500 µm are present in the samples. They evolve throughout the entire sample (Figure 5.2). The sample stored for 4 weeks at 20 °C has a layer of 300 µm at the surface which is very homogeneous without any cavities. This could be because of post-crystallization. However, due to limited imaging time at the synchrotron beamline only one sample for each storage and crystallization condition was imaged and thus no statistically valid conclusion regarding evolution over time can be made. The homogeneous surface layer could also be coincidence of sample preparation or due to molding process.

Air inclusion near the surface of untempered cocoa butter is not that pronounced as in case of tempered cocoa butter (Figure 5.1). But in contrast to tempered samples, large voids go through the entire sample. Thereby, spherical objects with a size of up

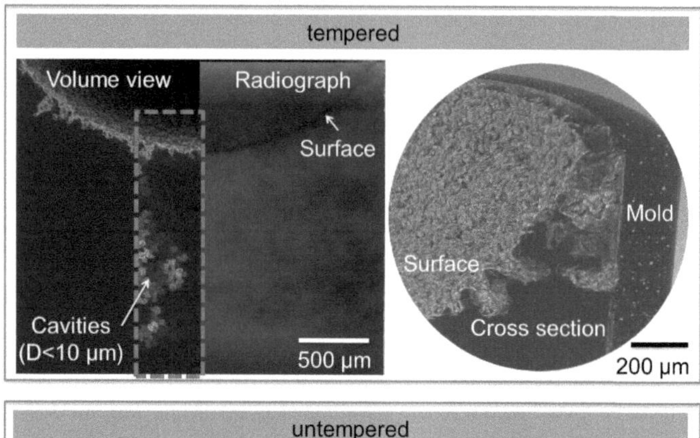

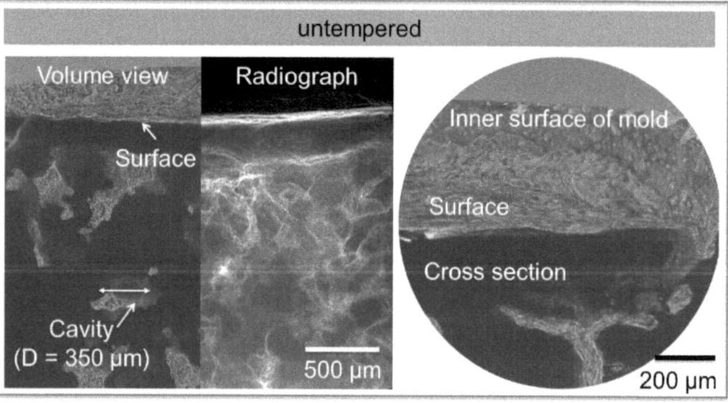

FIGURE 5.1: Volume view and radiograph of tempered and untempered cocoa butter presented as cross section and image of surface. Red rectangular selection shows voids which are locally present in selected samples of tempered cocoa butter. Samples stored for 4 months at 20 °C.

to 650 µm are present directly below the sample surface and within the sample of fresh untempered cocoa butter (Figure 5.2). The surface is rough with protrusion of up to 250 µm. Dahlenborg et al. (2011, 2012) imaged surface roughness of fresh tempered white chocolate pralines from a few microns up to 20 µm with confocal microscopy. This dimensions correlates with roughness at the top of cocoa butter (Figure 5.1). The tomography image clarifies that some channels evolve even further than only 20 µm into the sample volume.

The difference between tempered and untempered samples seen in the presented tomography images are also described in literature. Svanberg et al. (2011b) and Svanberg et al. (2011a) visualized microstructure development during solidification of pure cocoa butter and cocoa butter composites (cocoa powder or sucrose suspended in cocoa butter) with confocal laser scanning microscopy. They found substantial differences between samples being well tempered with seeds and samples not tempered with seeds. All samples tempered with seeding exhibited a homogeneous microstructure which formed within

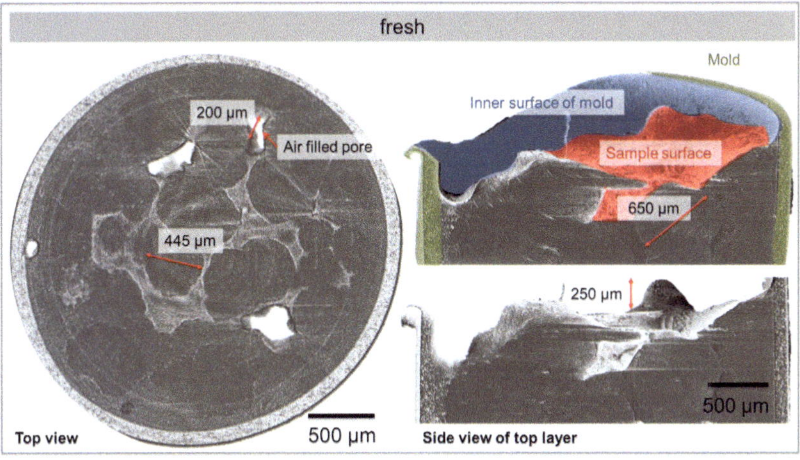

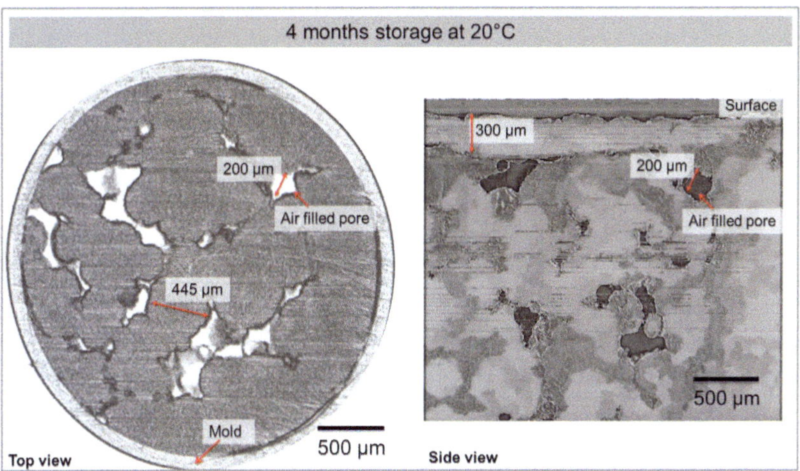

FIGURE 5.2: Cross section of untempered fresh cocoa butter and after storage of 4 weeks at 20 °C depicted as top and side view.

15 minutes. A small amount of liquid fat was found within the crystal structure. In contrast, samples without seeds showed a more heterogeneous structure with some large spherical crystals and large inclusions of liquid fat and small compact crystals. The non-seeded samples post-crystallized with storage time, so that after 1 week the amount of liquid fat was similar to the seeded samples. Wang et al. (2010) imaged fresh and stored untempered cocoa butter substitutes based compound chocolate. Topographic images and surface maps confirmed that fat crystals are homogeneously distributed on fresh chocolate. The surface of their samples appear in general smooth but some convexities and pits are present. Wang et al. (2010) proposed that this is a result of pores between fat crystals. On the contrary, bloomed chocolate has a marble like appearance according to Wang et al. (2010). Rousseau and Sonwai (2008) found ridges and

troughs on the surface of cocoa butter and chocolate samples which are amorphous and in the micron range with atomic force microscopy. The images are consistent with the tempered samples in Figures 5.1.

5.2 Investigation of change of cocoa butter microstructure due to addition of sucrose particles

Chocolate is a multicomponent system with particles (mainly sucrose, cocoa and in case of milk chocolate milk powder) embedded in a continuous fat matrix (mainly cocoa butter, microstructure shown in section 5.1 on page 71). Thus, chocolate microstructure is determined by the structure of the matrix phase and the arrangement and properties (such as particle size and shape) of the particle network. Cocoa butter structure forms during solidification from a liquid to a crystalline solid. It is likely that particles dispersed inside the solidifying fat phase alter the solidification process by e.g. heterogeneous nucleation or constraints of crystallization growth pathway (Rousseau and Sonwai, 2008). Therefore, chocolate model systems out of sucrose refined in cocoa butter with lecithin were imaged and analyzed.

Microstructure was analyzed for samples with and without precrystallization with synchrotron radiation. The dark spots in the tomography image of tempered samples are accumulations of pure cocoa butter from insufficient dissolution and mixing of the cocoa butter seeds (Figure 5.3). However, addition of seeds definitely enhance homogeneity of the cocoa butter based samples. In general, untempered samples show a looser structure characterized by more voids and fractal formation than tempered samples (Figure 5.3). In case of no precrystallization, pure cocoa butter exhibit cavities up to some hundred micrometers. They are smaller with addition of particles. Cocoa butter with a particles concentration of 1/3 by weight has small cavities of up to 50 µm. These are no longer detectable at higher particle concentrations of 2/3 by weight or in case of chocolate (Figure 5.3). Thus, cavities are smaller than the detection limit, which is about a micron in size. A lighter fractal like structure is visible around the cavities. This is probably because of edge enhancement effects from small void areas. Thus, small cavities are probably also present in untempered samples with higher particle concentrations. The lighter areas evolve through the entire sample. The voids possibly result from contraction of cocoa butter and the higher the particle content, the lower the total amount of contracting cocoa butter which leads to less porosity. Furthermore, the characteristic large spherical structures in untempered pure cocoa butter (Figure 5.1) are also visible with particle addition. However, the higher the particle concentration, the less pronounced are the large spherical objects (Figure 5.3). Thus, addition of more particles leads to a

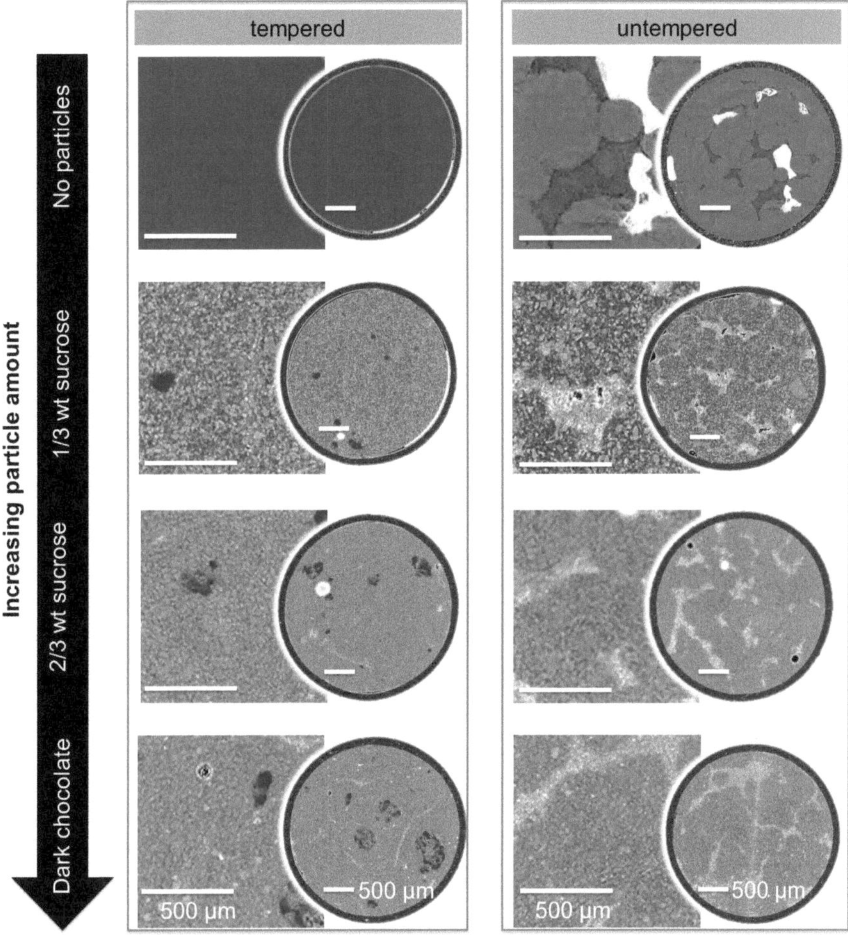

FIGURE 5.3: Structure of untempered and tempered fresh samples with variation in particle amount.

more homogeneous microstructure because the volume of cocoa butter, whose structure is significantly effected by the solidification conditions, compared to particles, which do not significantly change during the chocolate solidification process, is lower. Svanberg et al. (2011a) followed crystallization of cocoa butter samples with and without particles under the microscope. They observed differences in seeded and non-seeded samples with the latter ones resulting in large spherical crystals and a heterogeneous microstructure. With addition of sugar or cocoa particles they found enhanced crystal growth compared to pure cocoa butter in case of no tempering. A common explanation is heterogeneous crystallization, which however cannot be proven with Svanberg et al. (2011a) microscope images. The cocoa butter crystal network formed around sugar particles rather than starting from them. Svanberg et al. (2011b) and Svanberg et al. (2011a) found that in

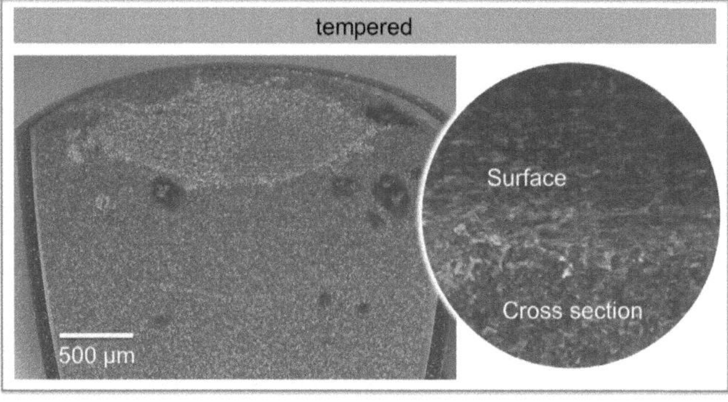

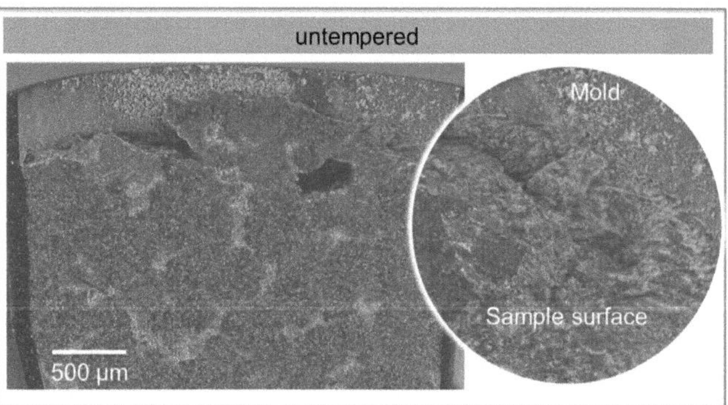

FIGURE 5.4: Fresh tempered and untempered sucrose refined in cocoa butter with lecithin (33 wt% sucrose) presented as cross section and image of surface.

case of addition of cocoa powder suspended in cocoa butter small rod-shaped crystals formed. They proposed that dissolved fatty acids, di- or triglycerides which are present in cocoa might act as seeds during crystallization. Sugar and cocoa particles had no major effect in case of tempering which can be confirmed with the tomography images (Figure 5.3).

The surface of cocoa butter with particles (33 wt% sucrose refined in cocoa butter and lecithin) is less homogeneous and smooth in case of no tempering compared to tempered samples (Figure 5.4). Surface roughness of tempered cocoa butter with sucrose samples is similar to that of pure cocoa butter. However, the surface roughness does not develop that far into the bulk material as in case of tempered pure cocoa butter. Rousseau and Sonwai (2008) study confirms these findings. They imaged the surface of cocoa butter and chocolate, which is basically cocoa butter with embedded particles, with atomic force microscopy in order to investigate the influence of dispersed particles within chocolate on cocoa butter structure and crystallization. They found lower surface roughness of fresh tempered chocolate compared to cocoa butter. A reason could be promotion of crystallization by the particles inside chocolate. This can be seen in Figure 5.4 as well.

With storage time growth of distinctive surface features were found in chocolate samples but not on the surface of pure cocoa butter which they attributed to the dispersed particulate network. Cocoa butter surfaces did not show any triglycerides fractioning and thus structural changes on cocoa butter were more homogeneous (Rousseau and Sonwai, 2008).

To analyze the impact of particle addition on microstructural changes with storage time, pure cocoa butter, aerated pure cocoa butter and cocoa butter with small sucrose (icing sugar, particle sizes between $d_{10,3} = 8$ µm and $d_{90,3} = 77$ µm) and large sucrose particles (household sucrose with particle sizes of $d_{10,3} = 255$ µm to $d_{90,3} = 470$ µm) were imaged with standard laboratory X-ray tomography. Small sucrose particles cannot be distinguished from the fat phase with standard laboratory tomography equipment (Figure 5.5 and 5.6), whereas large sucrose particles are identifiable. The density contrast can be enhanced with synchrotron radiation. However, only limited number of samples could be imaged with synchrotron radiation due to restriction in granted beamtime. Thus, to nevertheless use standard laboratory equipment to follow structural changes with storage time, an artificial capillary with a diameter of approximately 350 µm was molded into the samples. The capillary is large enough for detection. It was inserted vertically inside the cylindrical shaped sample. However, it was not possible to ensure same radial position for all samples over the total height. Furthermore, slight differences in position of the sample within the imaging area of the tomographic device lead to visualization of different positions. Thus, the capillary is not always at the same position in Figure 5.5. For determination of the diameter of the capillary, different positions are evaluated to capture the heterogeneity of the capillary dimension within the sample. Changes of diameter of the capillary can be measured (Figure 5.6). Addition of large sucrose particles shows the same behavior as pure cocoa butter. The capillary structure of both samples does not change during nine weeks of storage at room temperature but the diameter decreases when stored at elevated temperature. The void capillary diameter in aerated cocoa butter decreases with storage time and disappears when stored at elevated temperature. It is possible that a phase separation of air and cocoa butter took place and voids moved to the surface. This is strongly pronounced in case of storage at elevated temperature. Liquid content increases and viscosity decreases with rise in temperature probably facilitating phase separation. In contrast, addition of small sucrose particles seems to stabilize the structure. The capillary diameter does not change in these samples. The stabilizing effect of small particles might also explain the minor effect of storage time on the microstructure of tempered chocolate samples discussed in the following chapter. A reason could be that cocoa butter fat crystal structure is more dense and stable with small sucrose particles and thus the entire structure is stabilized as

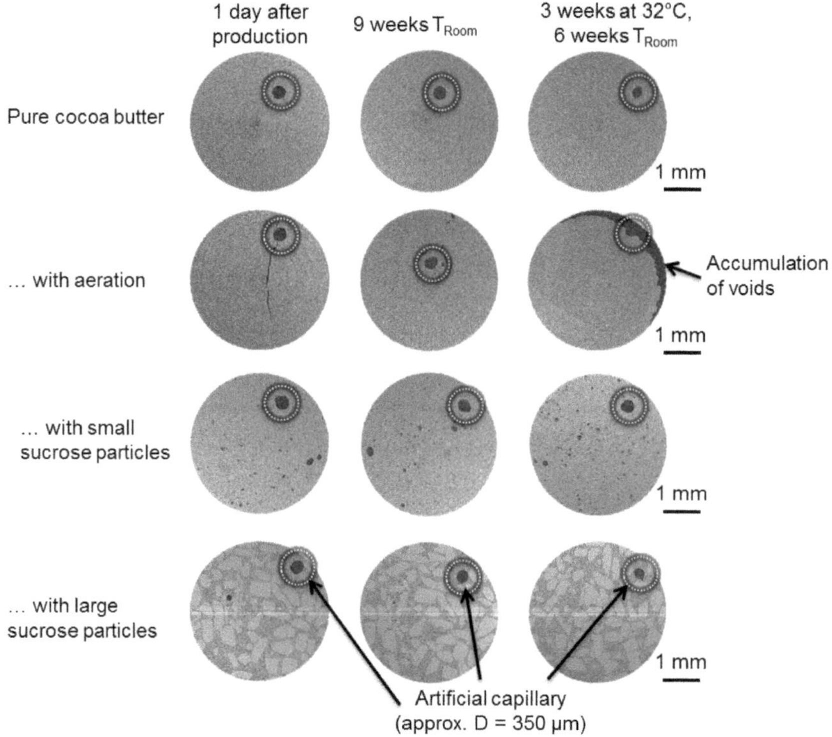

FIGURE 5.5: Cocoa butter samples a) pure, b) aerated, c) with 30 wt% small sucrose (icing sugar) and 30 wt% large sucrose (household sugar). Samples are imaged directly after preparation and after 9 weeks of storage at room temperature and elevated temperature. A capillary with a diameter of approximately 350 μm was added to follow microstructural changes.

discussed above. Furthermore, it could be that viscosity increases with particle addition which might lead to less change of structure even at higher temperatures.[1]

5.3   Visualization of the microstructure of conventional chocolate

The microstructure of industrial made dark chocolate was visualized with synchrotron radiation with a resolution of 1.175 μm per pixel. The tomographic images (e.g. Figure 5.7

---

[1]This experiment was only conducted once due to costs of tomography experiment. However, the results were confirmed with samples molded between two microscope slides and imaged under the light microscope. Thereby, significant changes in capillary diameter have been observed for pure and aerated cocoa butter but not for samples with small sucrose particles when samples were stored at elevated temperature. The diameter of the artificial capillary in samples stored at room temperature did not change.

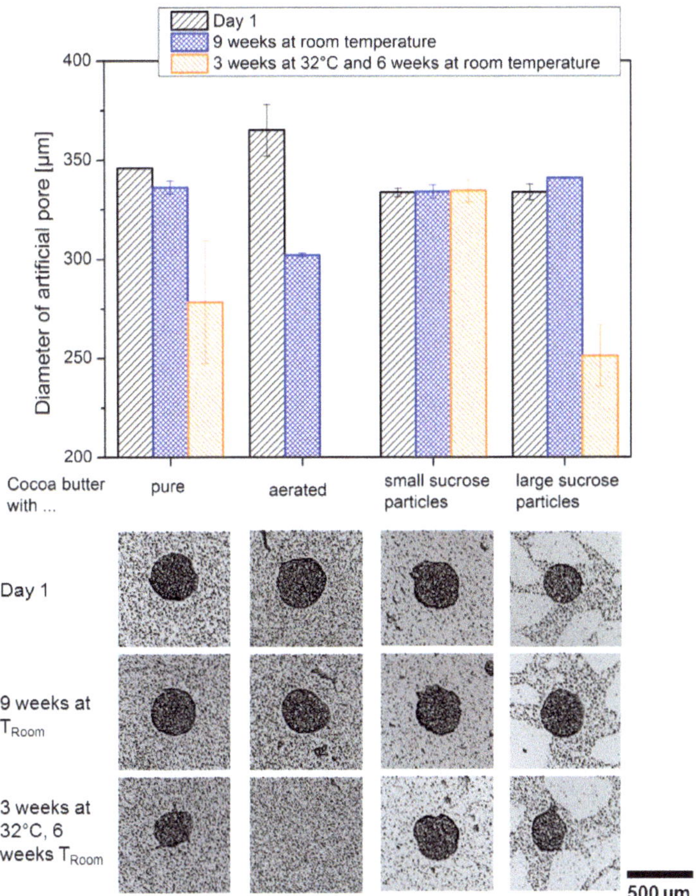

FIGURE 5.6: Change of diameter of artificial capillary in different samples. Cocoa butter samples a) pure, b) aerated, c) with 30 wt% small sucrose (icing sugar) and 30 wt% large sucrose (household sugar). Samples are imaged directly after preparation and after 9 weeks of storage at room temperature and elevated temperature.

and Figure 5.8) show that the particles are closely packed with a tendency to arrange with flat surfaces densely attached against each other. The particles have sharp edges induced by the refining process with a roller refiner. A distinction of separate particles from its surrounding is possible with a minimum size of approximately 5 x 5 pixels (about $\pm$ 3 pixel dependent on contrast). This corresponds to a minimum detectable particle size of 6 µm. The maximum particle diameter based on visual analysis of the tomography images is approximately 35 µm with only a few larger particles (e.g. see Figure 5.7 and 5.8). Larger particles such as shown in Figure 5.8 are the exception. The average size of most of the particles throughout the entire sample is about 10 to 25 µm (Figure 5.7). The size of the particles are distributed between a minimum and

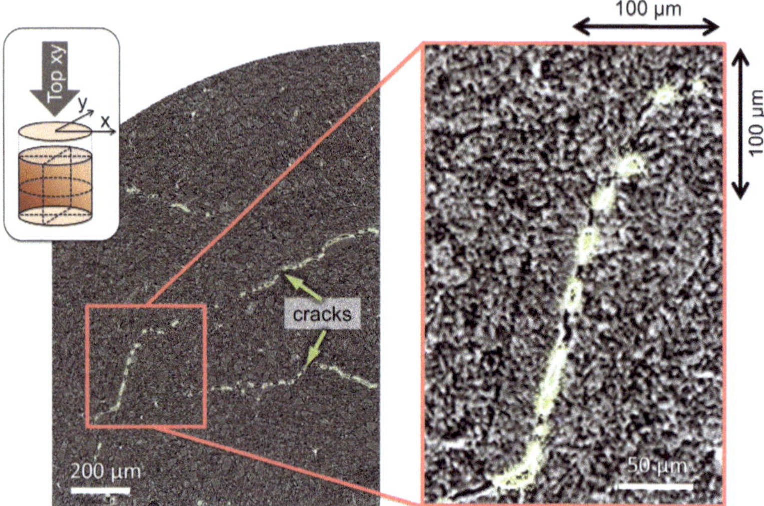

FIGURE 5.7: Image of chocolate sample, top view (cross section in the xy plane). Due to the partially coherent X-ray beam an edge enhancement effect is introduced to the tomographical reconstruction which emphasizes particle or crack borders within the tomogram. Cracks are highlighted.

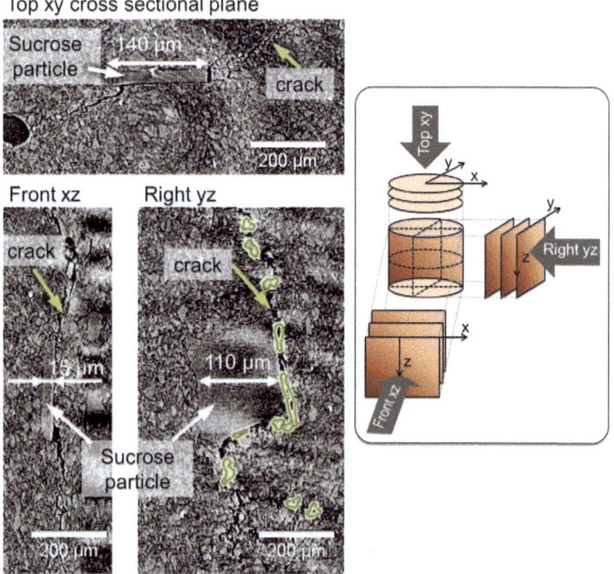

FIGURE 5.8: Cross sectional cuts at different planes within the sample showing crack in presence of a particle. Cracks are highlighted and Feret diameters of particles are given.

maximum size. This also corresponds to the given average particle size for the dark chocolate sample, which is $x_{50,3} = 10$ microns and a $x_{90,3} \approx 38\,\mu m$ (analyzed with laser diffraction) considering that larger particle are better evaluable from the tomography images and that the minimum particle size, which can be detected from the images is approximately 6 µm.

Furthermore, the microtomography images reveal voids in the solid chocolate with a diameter of up to 125 µm and cracks up to 7 µm and a length in the millimeter range. The voids and cracks seem to be distributed quite uniformly within the sample (see Figures 5.7) and can be found inside the sample center as well as close to the outer sample surface (see Figure 5.7). The cross section of the cracks is rectangular and elongated. It appears that they are primarily initiated at weak points within the chocolate sample, namely voids and particle boundaries. Figure 5.8 shows a crack evolving from a large sugar crystal in three different cross sectioned planes. The corresponding cross sectional planes xy from the top, xz from the front and yz from the right are illustrated in the sketch in Figure 5.8. The large particle is elongated in one dimension probably due to insufficient refining in the five-roll refiner in all three dimensions during industrial chocolate mass production. The particle is rectangular shaped and has a length of up to 140 µm in one dimension (seen from the top in Figure 5.8). This particle is disproportionately large and thus an exception. It cannot be seen as a representative particle inside chocolate. It seems that the crack starts to propagate from the boundary surface between cocoa butter and the particle or within the fat phase. The first mechanism is intergranular crack formation. Transgranular propagation with cracks through particles has not been observed and is very unlikely. It has to be evaluated whether cohesion forces within the fat phase or adhesion forces between particles and fat phase are weaker. However, it is more likely that cracks form at the interface, because the cocoa butter crystal structure is considerable stable. Furthermore, the cracks are mainly in circumferential hoop (tangential) direction. This is probably due to radial stresses (Figure 5.7) during the solidification process, which evolves from the outer surface to the interior of the sample. This induces contraction of the solidified front from outside to the inside of the sample and thus radial stresses.[2]

Untempered chocolate samples are characterized by a less uniform structure with less dense parts which prolong throughout the samples in a vein like structure (Figure 5.9[3], top right). These areas are less pronounced with storage time.

---

[2] Analysis of pure tempered cocoa butter samples without particles (Figure 5.3) showed no cracks. It seems like particles promote crack formation inside chocolate because more cracks are visible with increasing particle content (Figure 5.3).

[3] There are denser fractal parts with sizes of about 500 µm in tempered chocolate (Figure 5.9, middle left). These are undissolved cocoa butter seeds.

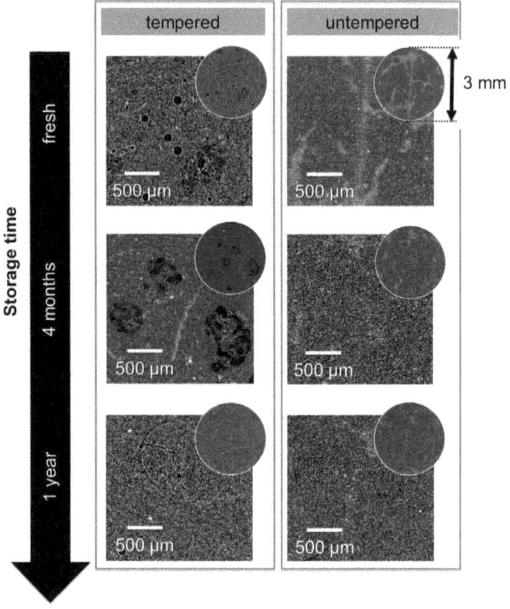

FIGURE 5.9: Structure of tempered and untempered chocolate directly after preparation (fresh) and after 4 months of storage at 20 °C and 1 year stored at room temperature.

5.4   Simulation of stresses due to solidification of the continuous lipid phase in a multicomponent system leading to crack formation

The particles are modeled as rigid bodies, which do not significantly contract due to cooling as they remain solid in the temperature range of the chocolate manufacturing process[4]. The matrix is modeled as a soft material, which contracts up to about $6 \cdot 10^4$ times more due to a decrease in temperature of 1 Kelvin. The rigid particles hinder the matrix to contract, which results in compressive stresses on particles and tensile stresses in the matrix. Figure 5.10 and 5.11 depict the stress distribution from the simulation due to a decrease in temperature from 25 °C to 10 °C, which are typical cooling conditions of chocolate. Chocolate is cooled from about 50 and 60 °C to 10 °C during production. However, in the range of 50 °C to 25 °C a significant amount of the cocoa butter is still liquid, so that stresses can be immediately relieved by movement of the liquid. At a temperature of around 25 °C cocoa butter changes from a more viscous liquid to form a solid network. Stresses grow with decreasing temperature (Figure 5.11).

---

[4]The thermal expansion coefficient of the particles was varied between $\alpha_{particle} = 10^{-4} \ 1/K$, $\alpha_{particle} = 10^{-5} \ 1/K$ and $\alpha_{particle} = 10^{-7} \ 1/K$, which are typical coefficients for solid materials. The results are shown in Figure 5.13 on page 85.

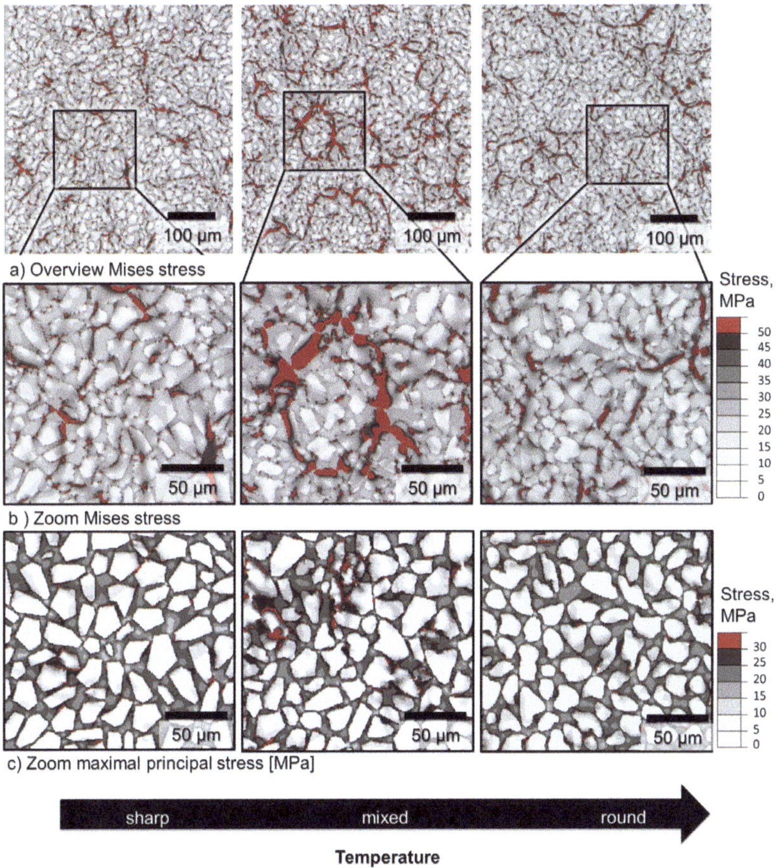

FIGURE 5.10: Simulated stress distributions due to temperature decrease from 25 °C to 10 °C of three model system composed of only sharp, mixed (sharp and rounded) as well as only rounded rigid particles embedded in a continuous soft phase, which significantly contracts. Area fraction of matrix phase is 42 %. Average Feret diameter of the particles is 23 µm.

The principal stresses are highest in the fat phase at the surface area of the particles (Figure 5.10c and 5.11), where they reach values of more than 30 MPa. Increasing stress was observed in regions with sharp edges and at the interface of particle and fat phase. Maximum principal stresses are highest where particles are closely packed (Figure 5.10c).

The stress state of a plastic material can be described with the von Mises stresses. They represent the multidimensional stress state of a plastic material in an uniaxial value. The von Mises stress distribution follows pathways within the sample (Figure 5.10a and b) and in case of the model in Figure 5.11 stresses are in a branched way from the large particle creating stress paths throughout the sample (Figure 5.11 and

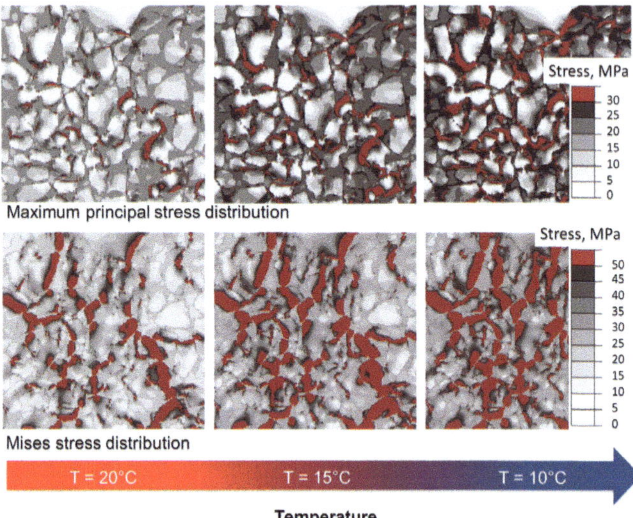

FIGURE 5.11: Simulated stress distributions due to temperature decrease from 25 °C to 10 °C in steps of 5 °C of a model system composed of rigid particles embedded in a continuous soft phase, which significantly contracts. Area fraction of matrix phase is 35 %.

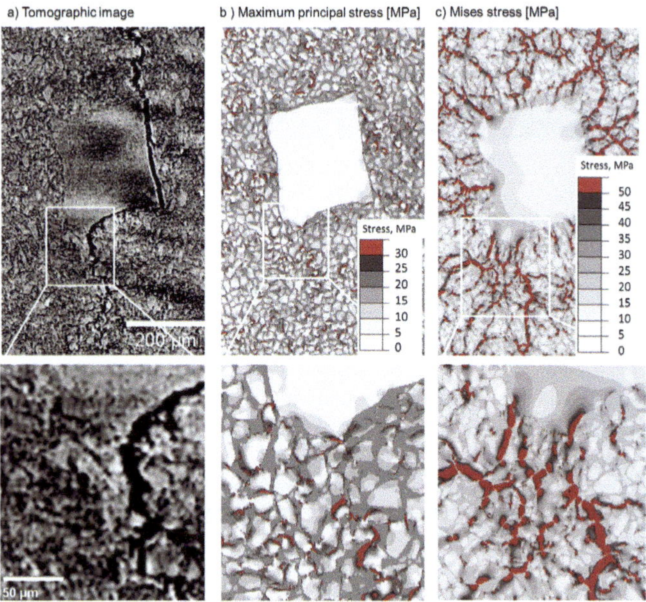

FIGURE 5.12: Maximum principal and von Mises stress distribution at 20 °C due to temperature decrease of 5 °C of a model system composed of rigid particles embedded in a continuous soft phase, which significantly contracts. The geometrical structure of the model system is based on the inner structure of chocolate as described by microtomography (a is a zoom of Figure 5.8). b) shows the maximum principal stress and c) the von Mises stress distribution. Area fraction of matrix phase is 35 %.

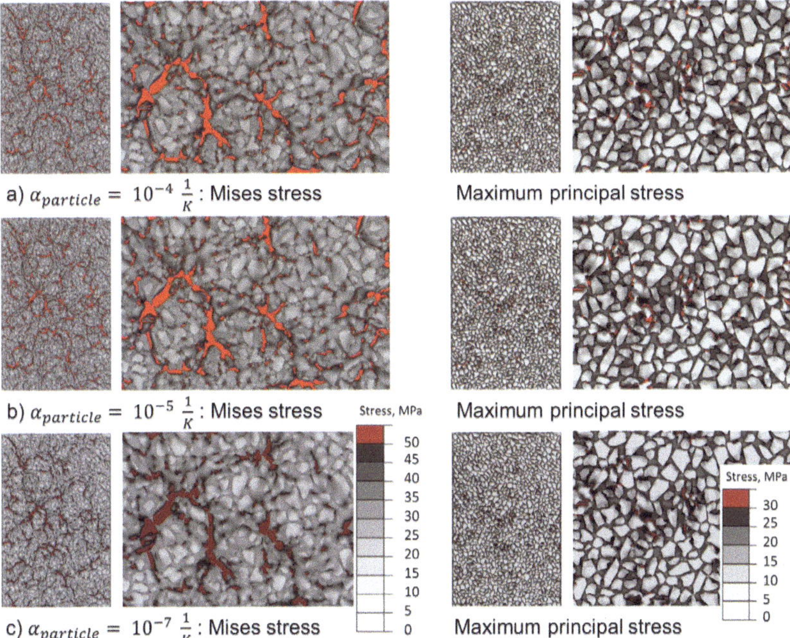

FIGURE 5.13: Maximum principal and von Mises stress distribution at 10 °C due to temperature decrease of 5 °C of a model system composed of rigid particles with a thermal coefficient of a) $\alpha_{particle} = 10^{-4}$ $1/K$, b) $\alpha_{particle}10^{-5}$ $1/K$ and c) $\alpha_{particle}10^{-7}$ $1/K$ embedded in a continuous soft phase, which significantly contracts.

Figure 5.12). The thermal expansion coefficient of the embedded particles is considerable smaller than that of the matrix phase in the investigated temperature range. So, changes of thermal expansion coefficient from $\alpha_{particle} = 10^{-4}$ $1/K$ to $\alpha_{particle} = 10^{-7}$ $1/K$ do not significantly influence the stress state after solidification (Figure 5.13). The reason is that stresses are mainly induced due to significant contraction of cocoa butter, which is in all cases much higher than that of the particles which is expected to be at ranges typical for solid materials which do not significantly contract such as glass or graphite and is typically in the range of $\alpha_{particle} = 10^{-4}$ $1/K$ to $\alpha_{particle} = 10^{-7}$ $1/K$ (e.g. Bailey and Yates, 1970, Cverna, 2002, DIN ISO 7991, 1998, Otto and Thomas, 1963).

The breakage force of cocoa butter at 20 °C was measured by Maleky and Marangoni (2011b). Resulting stresses for the given geometry are about 6 to 18 MPa. The published breakage force varies dependent on the crystal microstructure (Maleky and Marangoni, 2011b). Von Mises stresses exceeding the yield stress of the material results in plastic deformations, which lead to interfacial cracks if they exceed the material strength. Breakage of progressively brittle matrix may be caused by principal stresses. Both stresses develop values higher than the measured strength of cocoa butter at 20 °C.

The comparison with the tomographic image of a real chocolate system confirms crack formation at locations with stresses more than the measured breakage stress by Maleky and Marangoni (2011b) (Figure 5.12). Crack formation leads to a stress release. Thus, the chocolate likely breaks at its weakest point leading to stress release so that not all areas with higher stresses in the simulation are likely to finally fail. The simulation does not consider how stresses redistribute after the point of failure.

To conclude, this simple model of a two phase system with rigid particles embedded in a continuous soft phase with significant differences in contraction due to temperature decrease and crystallization could explain crack formation in chocolate. As an additional phenomenon solidification of the outer layer hinders further contraction of the entire sample during crystallization. This may lead to additional stresses within the sample and explains the preferential location of cracks in circumferential direction. For instance, Loisel et al. (1997) suggested that the contraction results in voids within the chocolate structure. However, pure tempered cocoa butter shows no detectable cracks nor voids. This might indicate that stresses induced by different contraction of particles and surrounding fat matrix are higher than the ones induced by a distribution of solidification front from outside to inside in radial direction.

Untempered samples contract less during initial solidification phase (Mehrle, 2007). However, they show significant post-crystallization (Rønholt et al., 2013) . This might lead to void formation throughout the entire sample as seen in Figure 5.2, which is possible the area around spherical crystallites.

5.5   Discussion of possible migration pathways and mechanisms based on visualized microstructure

The aim of the present study was to visualize the inner structure of chocolate to identify possible pathways for lipid migration possibly leading to fat blooming. Hartel (1999) as well as Smith and Dahlman (2005) suggested that small cracks and crevices present in chocolate might act as routes for lipid migration and that fat bloom seems to start at the pore outlet at the surface. Such cracks have been visualized in this study without destruction of the samples for the first time. Proper reconstruction of the images proved that cracks do not originate from the imaging technique. Sample destruction during the measurement would result in smeared images which was not observed. Convective flow of liquid cocoa butter fractions as a migration mechanism in chocolate was also proposed by Dahlenborg et al. (2011) and Aguilera et al. (2004). At room temperature about 25 % of the cocoa butter is liquid (Figure 2.5 at page 16). Aguilera et al. (2004)

specified that liquid fractions of chocolate fat might move through the porous network under capillary forces. Thus, the imaged cracks and voids within the microstructure of chocolate may have an important role in migration of lipids in chocolate.

Different mechanisms for migration through cracks and porosity in chocolate are proposed in literature. Altimiras et al. (2007) and Loisel et al. (1997) mentioned that the volume change when cocoa butter melts is forcing the liquid lipids to the surface through pores and cracks. Furthermore, Altimiras et al. (2007) experimentally found a linear relationship between fat migration and fat blooming.[5] And Dahlenborg et al. (2011) investigated the surface of chocolate and stated that smooth protrusion out of non-crystallized lipids with a diameter of a few microns up to 20 µm are initially forming at the surface acting as crystal nucleation points leading to fat blooming. Thus, Dahlenborg et al. (2011) proposed that fat bloom formation is a process were first pores transform to protrusions filled with liquid lipids with subsequent crystallization. They suggested convective flow as migration mechanism caused by a pressure gradient due to non-uniform contraction within the sample during solidification. This hypothesis is strengthened by the presence of cracks and pores at the surface (Figure 5.1) and throughout the entire chocolate sample (Figure 5.4 and 5.9) and the scanning electron microscopy images of bloomed chocolate by James and Smith (2009). They proposed that fat is extruded from the surface in a soft, but not molten state. The blade shaped crystals causing fat bloom seem to arise from cracks on the surface. It might be that liquid fat from low melting crystals migrate to the surface through the cracks or that solid soft crystals formed in the pores are extruded due to stress gradients and recrystallize without tempering and thus forming fat bloom. The cracks and voids found in chocolate may also explain the decreased bloom formation, which Hartel (1999) measured when replacing the dispersed crystalline with amorphous sugar particles. He observed an impact of the microstructure of dispersed sugar particles on chocolate fat bloom formation. He found out that amorphous sucrose leads to a reduced fat bloom formation compared to crystalline sucrose. He suggested that it might be due to a more closely packing of round shaped particles and thus a reduction in lipid migration rate or due to different geometry at the surface, which leads to different crystal structures. However, the microtomography images show that the particles in the current sample are very densely packed even though they are irregularly shaped and not round. Nevertheless, the difference in fat bloom formation due to changing amorphous to crystalline sucrose could be also explained with more lipid migration due to larger cracks. Cracks might preferably form in the presence of crystalline sucrose because of increasing local stress peaks in the presence of sharp edges of rigid particles. On the contrary, amorphous sucrose is viscoelastic in contrast to

---

[5]They determined fat blooming based on color change, in chocolate model systems consisting of sand particles suspended in cocoa butter. The fat migration was gravimetrically measured as the mass change of a filter paper stored underneath the sample.

crystalline sucrose. The viscoelasticity depends on moisture and temperature (Palzer, 2005). Thus, amorphous sucrose could follow the deformation of the continuous fat phase, whereas crystalline sucrose does not significantly deform. Consequently, there are reduced stress gradients between particle and fat phase. The lack of any cracks in tempered pure cocoa butter samples supports this hypothesis. Hartel (1999) concludes that the effect of the structure of the dispersed phase on bloom formation needs to be further investigated. Based on the presented tomographic images (Figure 5.1 to 5.9), it is proposed that differences can be explained with a variation of microstructure due to crack formation. The images show crack formation within the chocolate sample. Cracks are most likely initiated due to local stress gradients during solidification. Local stress gradients are induced by different volume changes of fat matrix and dispersed particles during cooling. The cocoa butter matrix contracts significantly more due to crystallization, ongoing crystal densification and thermal induced volume change while no or very little contraction of particles, especially sucrose crystals, occurs during solidification. This is expected to lead to stress gradients in the matrix and interface of cocoa butter and particles resulting in detachment of fat phase from particles and cracks. A simple model of a material with suspended particles in a fat phase based on a mixture of sucrose and cocoa butter could explain crack formation during solidification. Stresses within the material are in the order of magnitude of the strength of the fat phase. Stresses are higher in regions with many particles closely packed. Further research is required to determine if the found cracks act as the major migration pathway for lipids in chocolate causing fat blooming[6].

---

[6]Simulations in Chapter 6 show that relevant sizes of migration pathways in cocoa butter and chocolate are in the range of nm. Structures of less than about 4 µm cannot be identified with the presented X-ray tomography study.

6

Analysis of transport mechanism by visual observation of macroscopic migration in crystalline fat suspensions

The mechanism of lipid migration in chocolate is not fully understood and further mass transfer experiments of lipid transport in chocolate are needed (e.g. Quevedo et al., 2005). The aim of this study is to investigate the influence of the chocolate structure on migration on a macroscopic level and to investigate the transport mechanism. By understanding in how far structure influences migration, and by that possibly chocolate bloom formation, ways to manufacture products with better resistant against loss of quality can be developed (Hartel, 1999). The developed experiment is low in cost, simple to conduct and does not involve any hazardous materials and thus allows for easy screening of multiple samples. Model systems are used instead of conventional chocolate to reduce the complexity and to identify the major structural factors for migration rates in chocolate. Three different levels of simplification were applied. First, migration through porous glass was investigated as the basic model (system A). The properties of the porous glass are well characterized (e.g. Grüner et al., 2016) and the material does not change during the observation time. In addition, the pore space of the material is filled with air. It is not fully known if the structure of chocolate and cocoa butter consist of air filled pores, which would enable capillary rise, or if pores are filled with liquid lipids, which would mean that transport takes place through a liquid phase. The second model level (system B) is observation of migration from a highly oily filling through cocoa butter and chocolate samples. The experimental set-up of system B is based on Marty et al. (2005). Thus, the lipid phase from which the observed migration starts, has a high amount of mobile lipids, which can be transported through the more solid matrices of cocoa butter and chocolate. However, the structure of the matrix (e.g. pore space size, tortuosity, amount of liquid and air filled pores) is less characterized than the glass sample (system A) and might in addition alter during observation time due to crystal transformation or dissolution and recrystallization. The third level (system C) is migration from a cocoa butter based layer into another cocoa butter based layer firmly attached to the first one. Thus, the properties (such as liquid portion of lipids and matrix structure) of both phases are expected to be equal. The experimental system C was established to understand migration within plain chocolate rather than that from a highly oily filling usually described in literature (e.g. Guiheneuf et al., 1997, Khan and Rousseau, 2006, Marty et al., 2005, 2009, Miquel et al., 2001, Walter and Cornillon, 2002). The tempering degree of the 3 layers are altered in system D in order to analyze if difference between layers impact the transport. Cocoa butter, the matrix phase of chocolate, is used as the basic system and structure is altered by tempering process and addition of oil and particles. Migration rates are determined to compare experimental results to transport coefficients (e.g. diffusivity) reported in literature. In addition, migration is modeled with a capillary pressure driven approach to evaluate this model as a possible transport mechanism of lipids in chocolate.

6.1 Visualization of migration in cocoa butter samples

The experimental set-up is based on Marty et al. (2005, 2009). They stored cocoa butter on top of a model filling (palm kernel oil blend) stained with Nile red and detected migration of the colored front into cocoa butter samples with a flatbed scanner. Sudan red stain was used to visualize migration in this study. Sudan red is a group of oil soluble dyes, which are often used to visualize lipids (e.g. Bouchon et al., 2003, Dettmeyer, 2011, Patsioura et al., 2016, Qureshi et al., 2004).

To validate the experimental set-up of this study, migration from a stained sand and a sand with 30 wt% sunflower oil base into tempered cocoa butter (similar set-up to that of Marty et al. (2005, 2009)) is analyzed first. There is no significant migration from a sand base stained with 1.6 wt% red dye into cocoa butter, which is placed on top of the base (Figure 6.1). However, migration is observed from a base with 30 wt% sunflower oil (Figure 6.5). Thus, migration only occurs in combination with oil and is not a result of pure dye migration. The question is whether the detectable dye migration is proportional to that of the lipids or if the observed migration is only a result of dye transport due to a difference in chemical potential of Sudan red color pigments and

FIGURE 6.1: Tempered cocoa butter sample on top of layer with Sudan red dye mixed with sand. a) Directly after samples preparation, b) after storage for 1 week at 20 °C.

not comparable to that of lipid migration. Marty et al. (2005, 2009) concluded in their study that tracking the position of a front highlighted with oil soluble dye (in their case Nile red) into cocoa butter is an appropriate method to quantify oil migration. They compared migration of Nile red stain with that of linoleic acid from a peanut butter with palm oil base into cocoa butter. They found good correlation between the transport rate of the oil soluble dye and oil migration. Migration heights measured in this study with use of Sudan red instead of Nile red as dye are in accordance with the results from Marty et al. (2005, 2009) (system B, Figure 6.5).

The experimental set-up was further developed in order to analyze migration in plain chocolate (system C, Figure 6.5) without difference in oil content such as present between shell and filling (system B, Figure 6.5). Migration from a stained cocoa butter layer colored with Sudan red dye into a pure cocoa butter layer is visually detectable with the naked eye (Figure 6.2 a). An image of the cross section of a layered cocoa butter sample after 7 weeks of storage at 20 °C from this study (Figure 6.2 b) shows that the migration front is homogeneously distributed over the entire cross section. The interface between the layers is still detectable.

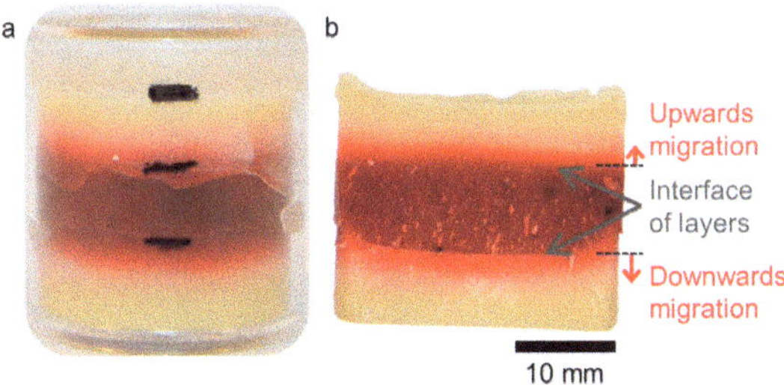

FIGURE 6.2: Image of sample with migration from a tempered cocoa butter layer stained with Sudan red into an unstained one. a) Tempered cocoa butter sample after storage for 7 weeks at 20 °C and b) its cross section. Middle layer is stained with Sudan red dye.

Migration from one cocoa butter layer to another one was also imaged with X-ray tomography (Figure 6.3 a, b and c). Lipiodol® was used to stain the bottom layer and detect migration into the top layer. Figure 6.3 a, a tomographic image of a fresh dark

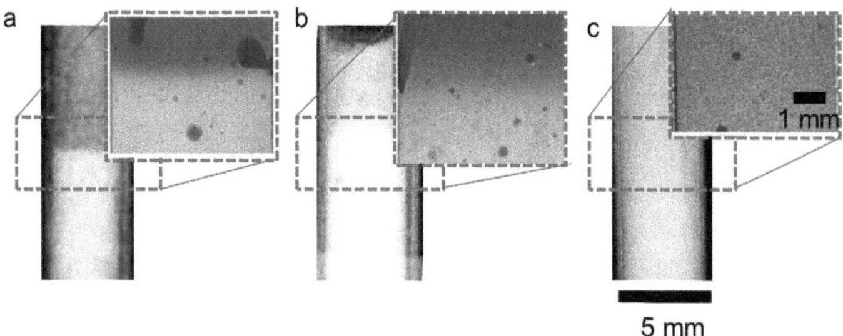

FIGURE 6.3: Migration from a tempered dark chocolate layer stained with Lipiodol® into an unstained one imaged with X-ray tomography. a) Fresh tempered dark chocolate sample with bottom layer stained with Lipiodol®, b) Layered tempered dark chocolate sample stored at 20 °C for 1 month, c) Layered tempered dark chocolate sample stored at 30 °C for 1 month.

chocolate sample, clearly shows two adjacent layers with the lower layer stained with Lipiodol®. After 1 month no clear distinction between the layers can be detected anymore (Figure 6.3 b and c). Storage at 30 °C lead to migration to maximum sample height, whereas migration of the sample stored at 20 °C did not fully spread until maximum sample height. The migration front is not precisely detectable with the tomography images in Figure 6.3. Nevertheless, observed upwards migration height with Sudan red was after 1 month at 20 °C about 1 mm and storage at 30 °C lead to migration throughout the entire sample height of 5 mm (Figure 6.5) which is consistent with the tomography images (Figure 6.3 b and c). Furthermore, the tomography images reveal that the layers are closely attached, which enables possible migration from one layer to the other. Insufficient attachment of both layers e.g. due to separation of the two layers during the solidification process prevents any migration from the stained to unstained layer. Consequently, migration cannot be observed. A possible explanation could be that capillary pressure is not sufficient for lipid transport if the distance is too large. The pressure drop rapidly decreases for larger capillary radii and higher contact angles (Equation 2.8 on page 28 and Figure 8.1 in the Appendix on page 136). Thus, the sample preparation can be used as a model to investigate migration within plain chocolate. The visual observation of the color front is used to discuss lipid migration in chocolate in the following. However, it can still not be fully excluded that migration of dye may not be proportional to that of lipids in chocolate. Nevertheless, the results of this study are overall consistent with literature findings determined with other experimental set-ups. Migration rates of the color pigments in comparison to that of lipids should be further investigated to elucidate if the rates and mechanisms of lipids correspond to the observed ones of the color pigments. Nevertheless, findings from color pigment migration, even if not equal to that of lipids, still provide information on the structure of the samples (e.g. changes over time, spatial distribution of porous structure or heterogeneities within the sample), which impacts both transport pathways and mechanisms.

6.2 Analysis of migration through white chocolate and porous glass cylinder

Small angle X-ray scattering revealed that lipids from liquid sunflower oil migrate into pores of chocolate powders in the size of 3.5 nm to 125 nm within seconds (Chapter 4). However, the investigated system contained with up to 10 vol% added liquid oil to the total powder sample volume a high amount of liquid oil as mobile phase compared to the powder into which migration was tracked. To further analyze the migration pathway and mechanism within a solid chocolate bar, a model system out of porous glass was brought into contact with a high oleic filling (system A). The structure (namely pore radius distribution, imbibition ability, volume porosity, tortuosity) of the porous glass was analyzed by Grüner et al. (2016), Lin et al. (1992). The pore size distribution has a median value of 3.4 nm. Furthermore, the lipids do not dissolve the structure and thus it can be expected that the matrix structure stays constant over the observation time.

Lipids from the palm-oil filling stained with Sudan red color migrate into the porous glass and migration height is linearly related to the square root of time after an induction period of about a day (Figure 6.4). Imbibition abilities of the 3.5 nm porous glass determined based on the detection of migration front from optical measurements are in the range of $\Gamma \approx 92.6 \pm 6.4$ x $10^{-7}\sqrt{m}$ for C14 and C12 at relative humidities between 24 % and 50 % for the lower bound and $\Gamma \approx 120.6 \pm 2.4$ x $10^{-7}\sqrt{m}$ for the upper bound (Grüner et al., 2016). The lower bound corresponds to a filling degree of about 45 % and the upper bound to a 100 % filling degree. The experimental results from this study are more comparable to that of the lower bound. The experimental data from imbibition of oil into the porous glass (Figure 6.4) gives an imbibition ability of $\Gamma \approx 94 \pm 13$ x $10^{-7}\sqrt{m}$ with the macroscopic properties of surface tension $\gamma = 35$ x 10^{-3} N/m, viscosity $\eta = 60$ x 10^{-3} Ns/m^2, a contact angle of $\theta = 19.3°$ (measured with sessile drop technique) and a volume porosity of $\phi = 0.3$ (measured value for initial volume porosity of Vycor porous glass from Grüner et al. (2016)). Thus, the imbibition ability, a characteristic parameter of the porous material is in accordance with the one from Grüner et al. (2016) and it can be consequently assumed that the observed transportation is similar to that of the investigated hydrocarbons from Grüner et al. (2016). Grüner et al. (2016) showed that transport through the glass can be described

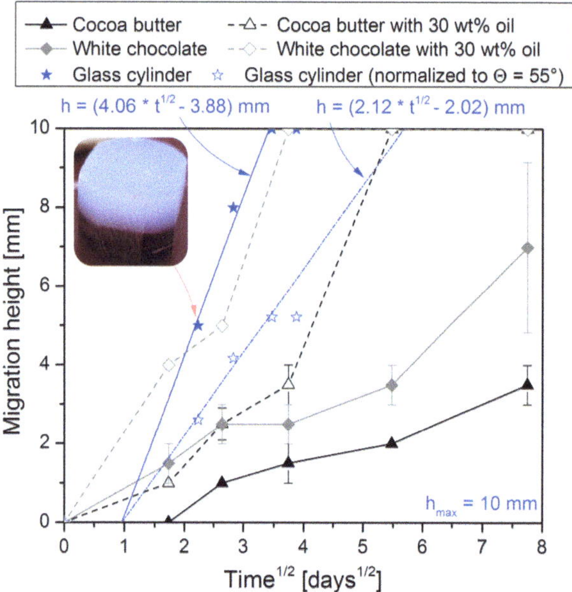

FIGURE 6.4: Migration height of oil-palmitin mixture into cocoa butter and white chocolate with no and with 30 wt% oil. Migration into glass cylinder with capillaries with radius of 3.5 nm. Normalized migration into glass from oil-glass contact angle of 19.3° to oil-cocoa butter contact angle of 55° is added.

with capillary rise[1]. The migration within the glass cylinder cannot be only a result of chemical potential differences of the dye because they have to be dissolved in oil to be transported through the porous glass (Figure 6.1 and Section 6.1 on page 90). Thus, they cannot be faster than the oil within the pores. And no unstained oil front was observed. The entire front was red (Figure 6.4, photograph in upper left corner).

The experimentally measured migration height into the porous glass cylinder was normalized in order to compare migration into the porous glass with that into cocoa butter and chocolate according to Equation 3.6 on page 52. Migration into porous glass (system A) is compared to that into white chocolate and pure cocoa butter (system B). Furthermore, 30 wt% of oil was added to the chocolate and cocoa butter to alter the structure. The normalized migration is close to the migration height of pure white chocolate and cocoa butter with oil during the first week and is intermediate to that of cocoa butter and white chocolate both with addition of 30 wt% oil after a week of storage (Figure 6.4). Maximum height of the porous glass cylinder was 10 mm and thus the last data point at day 15 has to be neglected in the discussion. It is likely that migration would have proceeded if sample height would have been larger. Migration through the porous

---

[1]Section 6.7.1 on page 121 shows the calculations of migration based on capillary pressure.

glass from day 1 until maximum sample height is linearly dependent on the square root of time, which is typical for capillary rise. In contrast, migration heights through chocolate and cocoa butter follow a stepwise manner with linear slopes in between temporarily stationery phases. Cocoa butter and chocolates are more complex than the glass cylinder and the structure of cocoa butter and chocolate changes over the observation time, which leads to variation of migration rates as observed in Figure 6.4. The migrating lipids may also dissolve part of the cocoa butter matrix as discussed in Chapter 4. Dissolution probably leads to changes of the structure over time. In addition, it is not known what effect the dissolution has on the fluid properties. The structure of the glass cylinder does not change and there are no effects of dissolution. Thus, migration heights are linearly dependent on the square root of time, which is typical for capillary rise and for diffusion. In contrast to glass, the structure of cocoa butter in its pure state or chocolate might change due to oil migration or recrystallization.

In contrast to the porous glass, the pores in chocolate and cocoa butter are probably (partly) filled with oil. Thus, there is a possibility of faster dye migration than oil migration. However, the experimental results are consistent with migration results from other studies (Figure 6.5). Furthermore, the geometric parameters such as pore size distribution and tortuosity are not that well characterized for cocoa butter. Porosity in the nanometer range in chocolates might be gaps in between crystal grains, nanoplatelets or crystal spherulites (Figure 2.2 in Chapter 2 on page 6).

Migration into pure cocoa butter is slowest, followed by migration into pure white chocolate. Migration in samples with addition of 30 wt% oil is fastest. Migration in white chocolate with 30 wt% oil reaches its maximum height of 10 mm after 2 weeks of storage on top of the highly oily filling (system B). Cocoa butter with oil addition showed migration up to 10 mm, which is the maximum height of the sample, from week 4 onward. Both samples without extra oil addition did not reach 10 mm migration height within observation time of 60 days. The stepwise manner and induction time in case of cocoa butter are discussed in the following Section 6.3 on page 98.

It can be seen that migration into white chocolate is faster than into pure cocoa butter. White chocolate consists of milk powder and sucrose particles embedded in a cocoa butter with milk fat matrix (Wohlmuth, 2009). Thus, increased migration rate could be either a result of the embedded particles[2] or because of higher amount of milk fat which contains more liquid and thus mobile lipids[3] or due to a combination of both. A significantly higher content of lower melting lipids might lead to a less dense structure

---

[2]Addition of sucrose particles to cocoa butter lead to slightly reduced migration (Figure 6.18). Thus, the impact of particles likely reduces the effect of milk addition, which might be even higher without any particles.

[3]In accordance to Figure 6.12 and Figure 6.13, addition of 30 wt% of sunflower oil accelerates migration (Figure 6.4).

and thus less resistance for transport and accelerated transport. This is consistent with the findings from Figure 6.5 discussed in the next Section 6.3.

6.3   Comparing migration height through tempered and untempered samples

Comparison with literature: Figure 6.5 compares measured data from this study with 2 different experimental set-ups (system B and C) to literature data. System B is comparable to that of Maleky and Marangoni (2011a) where a highly liquid filling migrated into pure cocoa butter. In system C on the other hand both layers (stained layer from which migration takes place and layer into which it occurs) are composed of pure cocoa butter and therefore the composition is the same except of the dye addition to one of the phases. The observed migration rates of this study are within the values published in literature by Maleky and Marangoni (2011a), Marty et al. (2005, 2009), Walter and Cornillon (2002). Walter and Cornillon (2002) investigated migration of lauric acid into commercial available dark chocolate whereas Maleky and Marangoni (2011b), Marty et al. (2005, 2009) used a fat-based model filling from palm kernel oil blend as stained phase and investigated migration into cocoa butter. Marty et al. (2005, 2009) used different thresholds for their image analysis for the height determination. It can be seen that the determined migration heights for saturation images and the 10 % value of dye intensity is higher than that of the hue images and the 50 % intensity threshold (Marty et al., 2009) (Figure 6.5). This indicates that the migration front is distributed over a range of up to 3 mm. Some migration (10 % intensity value) is faster than the majority (50 % intensity value). Migration heights in the present study were visually determined at the distance with significantly higher red color intensity than the unstained cocoa butter and thus rather correspond to the 50 % intensity value of migration front. There is good agreement between data from Marty et al. (2005, 2009) and migration heights of system B (Figure 6.5). Maleky and Marangoni (2011b) used the 10 % intensity value as threshold and thus migration heights are larger than the ones determined in this study.

Comparison of migration in tempered and untempered cocoa butter: In accordance with Marty et al. (2005, 2009), migration in untempered cocoa butter is faster than in tempered samples (Figure 6.5). Maleky and Marangoni (2011b) investigated migration in cocoa butter samples which have been crystallized in three different ways (static, sheared and oriented, Figure 6.5). They found that migration in dense polycrystalline networks is slower than in samples with larger crystallite sizes. They suggest that structures in that size retard the oil transport. The crystalline network in untempered

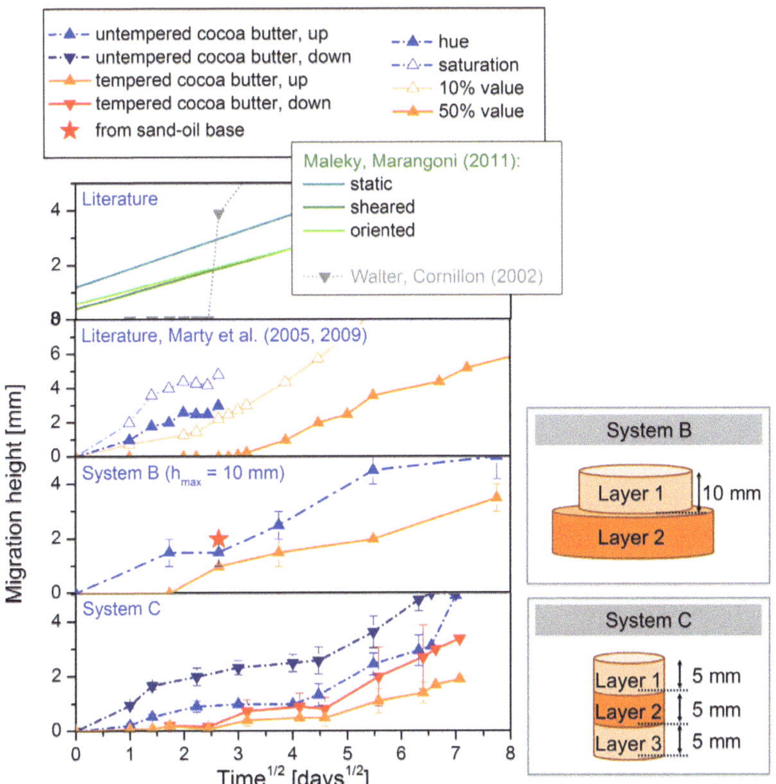

FIGURE 6.5: Macroscopic migration into tempered and untempered cocoa butter. System B consist of tempered or untempered cocoa butter on top of a Palmin blended with sunflower oil and Sudan red (0.69/0.30/0.01 by weight) and tempered cocoa butter on top of a sand blended with sunflower oil and Sudan red (0.69/0.30/0.01 by weight) base. System C are 3 layers comprised of tempered or untempered cocoa butter with the center layer stained with Sudan red (concentration of 0.01 by weight). Literature data from Maleky and Marangoni (2011b), Marty et al. (2005, 2009), Walter and Cornillon (2002) are given for reference.

cocoa butter samples is less dense and more heterogeneous than that of tempered samples[4]. This leads to faster migration through untempered cocoa butter than tempered ones as seen in Figure 6.5.

The solid fat content of seeded and non-seeded samples on a macroscopic level is the same (Svanberg et al., 2011c). However, Svanberg et al. (2011c) found that the microstructure was significantly different resulting in different migration rates. In case of non-seeding, variance was higher than with seeding which can be explained with a higher heterogeneity of the microstructure. This also corresponds to the findings of this study. It was visually

---

[4]Chapter 5 (e.g. Figure 5.3 on page 75) discusses the different structures of untempered and tempered cocoa butter in more depth.

observed that the migration front is more sharp in case of tempering with seeds than without tempering (Figure 6.2 and Figure 6.21). In addition, migration rates varied more for untempered samples but were rather stable for tempered samples (Figure 6.23 on page 118 in Section 6.6). The experimental results suggest that it is important to obtain a dense microstructure at all times in order to reduce migration, which is in accordance to the conclusions of Svanberg et al. (2011c).

Comparison of migration from a highly liquid filling into cocoa butter and that within two cocoa butter layers: Samples from Maleky and Marangoni (2011a) and system B correspond to the faster values for downward migration of the samples in system C from this study although migration in system B and Maleky and Marangoni (2011a) was upwards. Migration in system C from layer 2 into layer 1 and 3 was measured in two directions (upwards and downwards). Migration heights are different in upwards and downwards direction. Migration into the lower layer 3 (thus layer which has been molded first) is faster than migration upwards into the layer which was molded on top of the other layer 1[5].

Induction time: All tempered cocoa butter samples from system B and C of this study and published data from Walter and Cornillon (2002) as well as Marty et al. (2005) exhibited an induction time without any detectable migration (Figure 6.5). The induction time is in between 3 days and 1 week. In contrast, migration in untempered cocoa butter samples show no induction time of migration height. Walter and Cornillon (2002) suggest that the induction time is a result of formation of a "particular internal structure of chocolate", which might be a prerequisite for migration, during the first days. Migration through porous glass (system A) also showed a lag phase of about a day. Thus, it might also be that the mobile phase undergoes a transition or needs to rearrange leading to a lag phase.

Stepwise migration: In general, migration height is not constant over the square root of time through cocoa butter but through porous glass (Figure 6.4 on page 94 in Section 6.2). Instead it seems like migration heights progress in a stepwise manner with time. First, an induction time for the first days with little or no migration is observable followed by an increase in migration height and another slow or no migration time period before another increase in migration towards the end of observation time. For example, migration height increases up to a week of storage time to a height of about half the maximum samples height ($h_{max} = 5$ mm) in untempered cocoa butter in downwards direction (system B, Figure 6.5). Between 1 and 2.5 weeks migration height is quite constant with less than 0.1 mm increase on average upwards and 0.5 mm downwards.

---

[5]A detailed analysis of direction of migration is presented in the following based on data from Figure 6.6.

Migration height then again increases up to the maximum migration height of 5 mm from 2.5 weeks of storage until the end of observation period which was 50 days. The overall migration rates, which are presented in Figure 6.23, were similar for untempered and tempered samples in system B and C of this study and Maleky and Marangoni (2011a) after the induction time. However, total migration height differed due to different induction times.

Change of migration rates with square root of time was also reported in literature and can be explained with structure changes. Marty et al. (2005) investigated a plateau in migration of an oily base into untempered cocoa butter (Figure 6.5) after about 4 days up to a week. However, another increase in migration height was observed at day 7. In accordance to Marty et al. (2005) and Figure 6.5, Hondoh et al. (2016) found that weight gain of chocolate with silicone and canola oil seems to have an equilibrium after a week with another increase at day 20. Khan and Rousseau (2006) measured the evolution of triolein to 1,2-distearoyl-3-oleoylglycerol, a main component of cocoa butter, over storage time for a composite of dark chocolate with hazelnut filling at 11, 20 and 26 °C with HPLC. In general, triolein content and thus migration of hazelnut filling into chocolate is higher at elevated temperatures. Furthermore, the development seems to follow a stepwise manner similar to that observed in Figure 6.19. At 26 °C a first equilibrium is reached after about 1 to 3 weeks. Nevertheless, another increase in triolein content can be seen after about 6.5 weeks of storage. At a temperature of 20 °C a first increase in triolein content is measured after about a week, which then stays quite constant and increased again after 2 weeks. Triolein content at 11 °C is very low up to end of observation which is 8 weeks. The stepwise migration might be a result of a change of structure of the crystalline network over time. The migration in stages corresponds to the evolution of the crystal network over storage time. Smith et al. (2007) measured an increase in amount of the polymorphic form β_{VI} within the solid phase of cocoa butter with varying amount of hazelnut oil content (0, 1, 5 and 20 wt%) over storage time at 20 °C and 25 °C. In general, an increase in polymorphic form β_{VI} is faster at higher oil contents and elevated temperature. Furthermore, it also seems like the increase follows a stepwise manner. For example, the first polymorphic form β_{VI} for pure cocoa butter stored at 25 °C was measured after 2.5 weeks, which corresponds to the observed induction times of this study, and reached a quasi constant value after 4 weeks up to 5 weeks with a subsequent additional increase. And with 20 wt% of hazelnut oil a first plateau is reached after 3.5 weeks with another sudden increase at 4 weeks of storage at 20 °C. van Malssen et al. (1999) reported that melted cocoa butter, which is not precrystallized, first crystallize into the α_{II} form at 18 °C within 1 hour (solid fat content is about 50 % after 1 hour at 18 °C) from which it converts into the β' (form III or IV) form and finally after 1 week to the β form. Wille and Lutton (1966) measured a stability of 3 hours (start of transformation) to a day for cocoa butter in form IV (transformation completed) and 7 weeks to more than 18 weeks for form V at 21 °C. In addition, Adam-Berret et al. (2011) investigated the evolution of a tristearin and tricaprin fat crystal network with NMR. They found a decrease in specific surface area of the fat crystal network which they explain with Ostwald ripening and thus melting of small crystals which recrystallized on the larger crystals. However, the tortuosity in their fat network did not change over measurement period of up to 2 weeks with higher tortuosities at high solid fat contents (2.3 at 75 % compared to 1.3 at 15 % solid fat). Afoakwa et al. (2009a) found that hardness of under-tempered dark chocolate increase during the first 3 to 5 days of storage when it reached a plateau until end of observation time, which was 7 days. This corresponds to the first step of migration, which reached its plateau after about 4 days holding it up to 2 weeks. The plateau in system B is shorter than that observed in system C (Figure 6.5). This is especially pronounced for untempered cocoa butter. It might be that the higher liquid content of the filling base from system B accelerates the structural changes and thus the steps follow each other faster with less time of migration height resting in a plateau.

Influence of layer position: One possible explanation for the differences between upwards and downwards migration could be the cooling time. The layers are molded subsequently with 30 minutes solidification time at cooling conditions prior to molding of the top layer to avoid mixing of stained and unstained layers during molding. Therefore, layer 3 (see Figure 6.6 for sketch of sample set-up) is cooled for 1.5 hours, layer 2 for 1 hour and layer 1 only for 0.5 hours. In order to investigate if cooling time has an impact on observed migration heights, different layers were stained and migration from different layers were compared. Downwards migration from the very top layer 1 (solidification for 0.5 h at 5 °C) into middle layer 2 (solidification for 1 h at 5 °C) and migration from the middle layer 2 into bottom layer 3 (solidification for 1.5 h at 5 °C) as well as upwards from layer 3 into 2 and from 2 into layer 1 are compared for tempered and untempered samples (Figure 6.6). The trend is the same for migration of both layers. Maximum height difference between two data points at a specific time is 1.1 mm downwards (with average difference of 0.3 mm) and 0.6 upwards (with average difference of 0.2 mm) for tempered samples and 1 mm downwards (average 0.3 mm) and 0.6 mm maximum difference in upwards direction (average 0.1 mm). Differences are within standard deviation and the detection limit of the experiment. Figure 6.6 supports the findings from Figure 6.5 with highest migration heights for untempered samples (downwards) and lowest for upwards migration in tempered samples. Furthermore, tempered samples have an induction time of 1 week before significant migration occurs. Migration in untempered samples can be detected from day 1 on. Untempered samples reach their maximum sample height of 5 mm after observation time of 50 days,

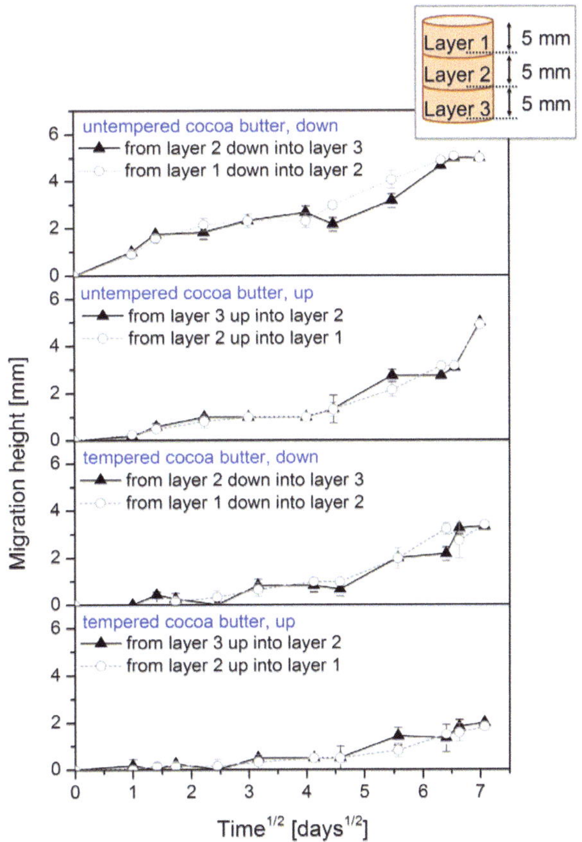

FIGURE 6.6: Macroscopic migration of tempered and untempered cocoa butter samples in upwards and downwards direction. Maximum sample height is 5 mm. Migration from different layers are compared. Storage of samples at a temperature of 20 °C.

whereas migration in tempered cocoa butter is only up to maximum 3 mm after 50 days of observation time. Consequently, migration from different layers is basically indifferent when direction and tempering is kept constant. Thus, only precrystallization and direction influence macroscopic migration in case of pure cocoa butter and not differences in cooling time between 0.5 and 1.5 hours.

Influence of gravity on migration: Besides cooling time, the faster migration in downwards direction might result from gravity. Thus, the influence of gravity on macroscopic migration in cocoa butter samples was investigated. Therefore, samples were stored in standard direction (bottom layer was layer 3 which was molded first) and upside down (bottom layer was layer 1 which was molded last as sketched on top of Figure 6.7). Results for samples with addition of 10 wt% of oil are presented in Figure 6.7. These findings have been reproduced with other compositions such as an addition of 5 wt% oil

stored in standard, upside down direction and turned to the side (Figure 6.8). Figure 6.7
and 6.8 show that gravity is not the reason for faster downwards migration from layer 2
into layer 3. There is not much migration with only maximum migration of 0.5 mm after

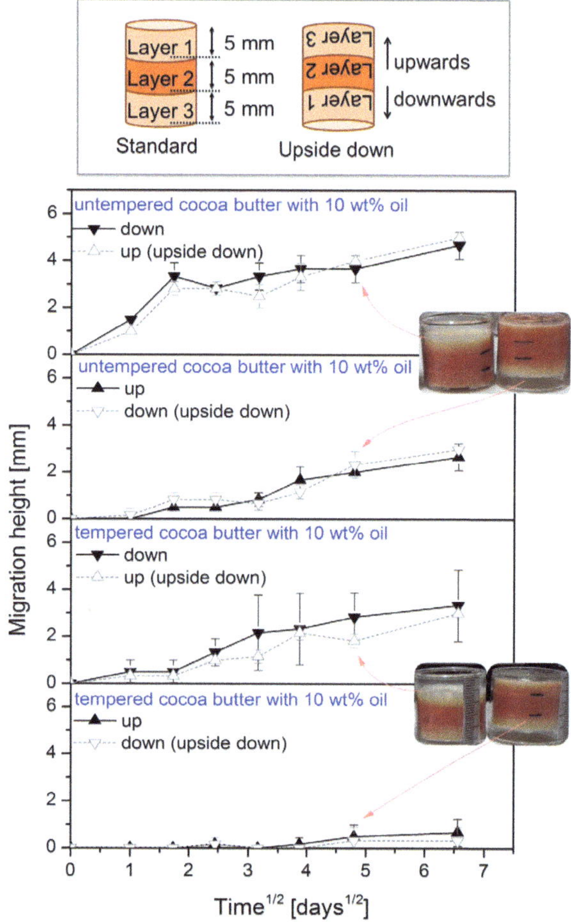

FIGURE 6.7: Macroscopic migration from tempered and untempered cocoa butter with
10 wt% oil layers upwards and downwards into layers of tempered and untempered
layers. Maximum sample height is 5 mm. Migration from samples with standard set-
up and samples stored upside-down are compared. Storage of samples at 20 °C.

maximum observation time of 43 days in tempered samples upwards with the standard
set-up and downwards with the upside down set-up. However, there is up to 3 mm average migration for tempered samples downwards in standard set-up and upwards when
samples are stored upside down. To sum up, migration downwards from a standard
sample is equal to migration upwards from a sample stored upside down. But there is
significant faster migration into a layer molded before the layer from which migration
takes place (downwards in standard set-up) compared to migration from a layer molded

prior to the layer into which migration occurs (upwards in standard set-up) even against gravity. Whereas migration into the same layers is equal for both set-ups with faster migration into layer 3 than layer 1 from the middle layer. Untempered samples show the same development for migration height with observation time.

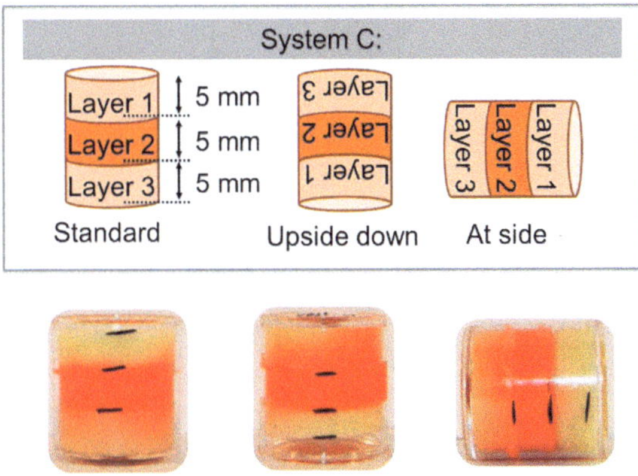

FIGURE 6.8: Images of layered cocoa butter with 5 wt% oil samples after 3 weeks stored at 30 °C. Stored in standard direction, upside down and on the side.

Influence of structure: If gravity (Figure 6.7) and cooling time (Figure 6.6) can be neglected as major impact factors for macroscopic migration, the reason for the direction dependence observed (Figure 6.5) has to be a result of differences in the interfacial region due to solidification against a cocoa butter surface compared to the interface which originally solidified in presence of air. The structure at the interface of two adjacent layers was analyzed with tomography with a focus on differences between top and bottom layer. Cross sections of the samples captured with tomography show differences at the interface between top and bottom layer (Figure 6.9, 6.10 and 6.11)[6]. The tomography images (Figure 6.9, 6.10 and 6.11) support the findings from Figure 6.6, 6.7 and 6.8 that the structure of the two layers close to the interface is different and consequently leads to different migration heights. The structure at the bottom layer (Figure 6.9 e) is very similar to the structure of the standard sample at the surface with air (Figure 6.9 b). The bottom layer of tempered cocoa butter samples is homogeneous without any cavities. Small cavities are directly at the interface up to 500 µm below the surface. In contrast, the top layer (Figure 6.9 d, Figure 6.10) is less homogeneous at the cocoa butter layer interface. Darker areas are visible (Figure 6.10). These areas are cocoa butter seeds used for tempering (Figure 6.10 and further discussion in Chapter 5 and Section 5.3

---

[6]The bottom layer was stained with Lipiodol® which is dissolved iodine in sunflower oil for tomography (not for migration samples). Thus, addition of oil might have lead to differences of the structure seen with tomography. However, for pure samples without Lipidol® shown in Figure 6.9 the imaged layer, which is close to the surface (Figure 6.9 b) is similar to the bottom layer adjacent to the interface of layered samples (Figure 6.9 e).

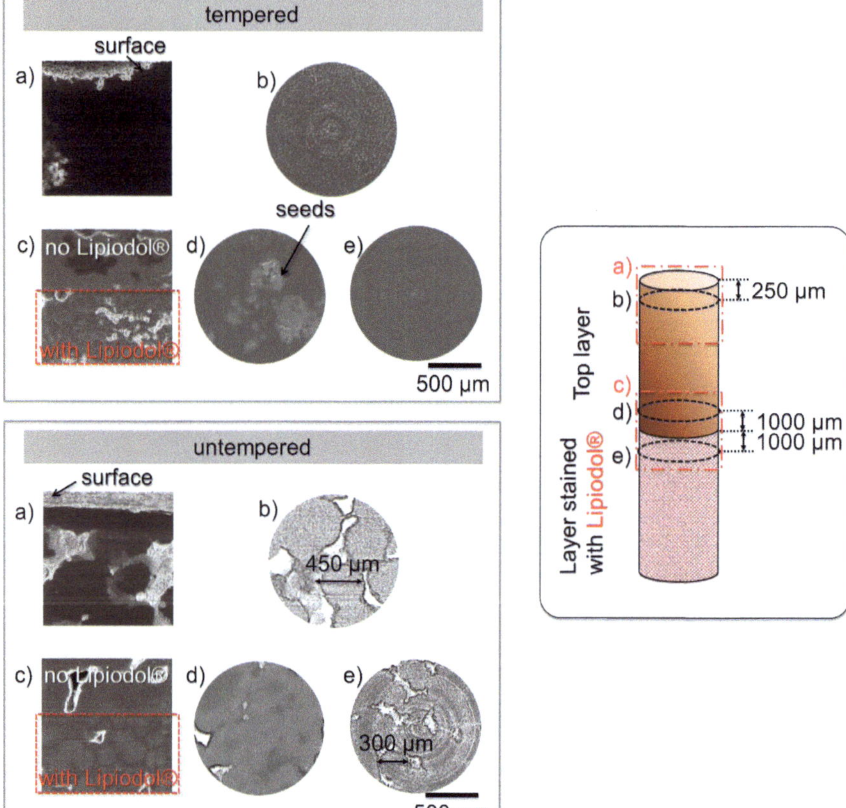

FIGURE 6.9: Tomography images of tempered and untempered cocoa butter. Surface of standard samples are shown. Images of layered samples with lower layer stained with Lipiodol® are given. Top views are given as circular cross sections in addition to side views.

on page 78) which can be seen only in the top layer of the tempered sample. The untempered layered cocoa butter sample has spherical objects with a size of about 300 to 450 µm close to the air surface (Figure 6.9 a, b) and at the bottom layer close to the interface (cross section e in Figure 6.9). Furthermore, the structure is characterized by voids with a size of about 100 µm. The spherical objects are not that pronounced in the top layer close to the interface (cross section d, Figure 6.9, Figure 6.11)[7].

Differences between upper and lower layer at the interface can be a result of different solidification of the surface layer in presence of air or because cocoa butter poured on top of an already solid cocoa butter crystallizes in a different way. In that case, the solid top layer of the lower cocoa butter layer might act as crystallization side for homogeneous

---

[7]Chapter 5 presents a detailed discussion of microstructure of tempered and untempered cocoa butter.

FIGURE 6.10: Tomography image of tempered cocoa butter as layered samples with lower layer stained with Lipiodol®.

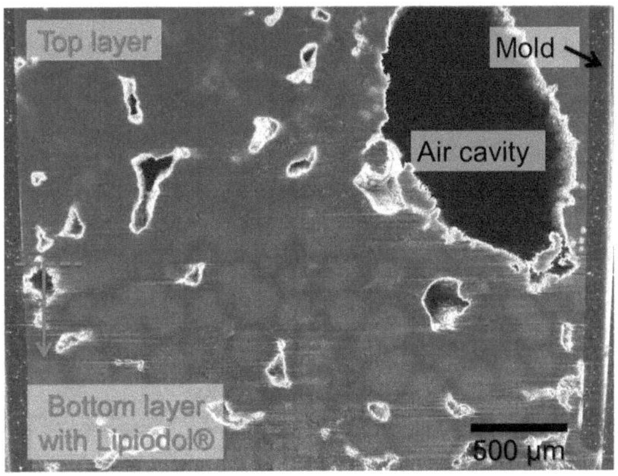

FIGURE 6.11: Tomography image of untempered cocoa butter as layered samples with lower layer stained with Lipiodol®.

crystallization of the cocoa butter poured on top with fast growth of the fat crystal network. Svanberg et al. (2011a) found that samples with cocoa butter seeds rapidly form a dense and homogeneous crystal network. A detailed analysis of structure of an entire sample could clarify that question.

This could also explain why upwards migration in system B and migration heights reported by Marty et al. (2005, 2009) are more similar to downwards migration of system C than the slower upwards migration (Figure 6.5).

6.4 Analysis of migration through lipid mixtures with varying oil content

Sunflower oil was added to cocoa butter in different ratios to analyze the impact of addition of low melting lipids to cocoa butter on macroscopic migration. Therefore, 5, 10, 20 and 30 wt% of oil have been added after tempering and prior to molding for system B[8]. Downwards migration at 20 °C and 30 °C is shown in Figure 6.12 and upwards migration in Figure 6.13. Samples with 5 and 20 wt% were only observed up to 4 weeks but are expected to follow the same trend from week 4 to week 8.

Migration is in general faster at 30°C than at 20 °C. This is in accordance with the results from Guiheneuf et al. (1997) who found substantial differences between the migration profile of a hazelnut filling into chocolate at 19 and 28 °C. Their system was composed of chocolate solidified on top of a hazelnut filling base. They measured slower migration at 19 °C than at 28 °C (data points from Guiheneuf et al. (1997) are added to Figure 6.18).

Addition of some sunflower oil to the cocoa butter fat phase leads to reduction of migration (Figure 6.12 and 6.13). For downwards migration addition of 20 wt% at 20 °C and 10 wt% at 30 °C results in migration heights equal or smaller than for pure cocoa butter (Figure 6.12). In case of upwards direction, 20 wt% or less of oil addition leads to migration heights of less or equal to that observed in pure cocoa butter samples at both storage temperature of 20 °C as well as 30 °C (Figure 6.13). The decrease of migration heights at addition of small amounts of oil, such as 5 wt% and 10 wt%, are possibly a result of different sample structures leading to reduced migration rates. Migration is in general influenced by the driving force such as a concentration gradient or pressure difference and resistance against migration, which is dependent on the matrix structure (Ziegler, 2007). The only difference in composition of all three layers within one sample is that of Sudan red color, which is only present in one of the three layers. Oil content was kept constant for all three layers of a single sample. Thus, concentration differences of color is kept constant for all samples and oil concentration is equal in all layers within one sample. Consequently, a change in driving force based on a concentration gradient

---

[8]10 wt% of oil was also investigated in order to analyze the direction of migration (Figure 6.7). The measured heights for downwards migration in standard set-up for that sub-study were higher than the ones presented in Figure 6.12. The two sub-studies have the same set-up but were prepared by different persons on different days. Sample preparation was done manually and thus stirring was kept constant for each experimenter and sub-study but may differ between the sub-studies. The standard deviation of Figure 6.7 for downwards migration are quite high which indicates that the triplicates were not all equal. In contrast, standard deviations of upwards migration and migration of this sub-study (Figure 6.13) are small and within detection limit. As Figures 6.7 as well as 6.13 suggest, the sub-study are both consistent within itself.

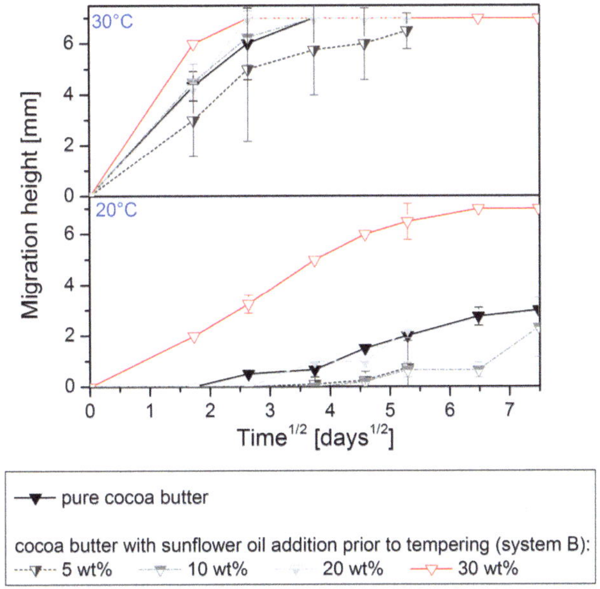

FIGURE 6.12: Downwards migration of tempered cocoa butter samples with addition of oil at different concentrations. Storage of samples at 20 °C and 30 °C.

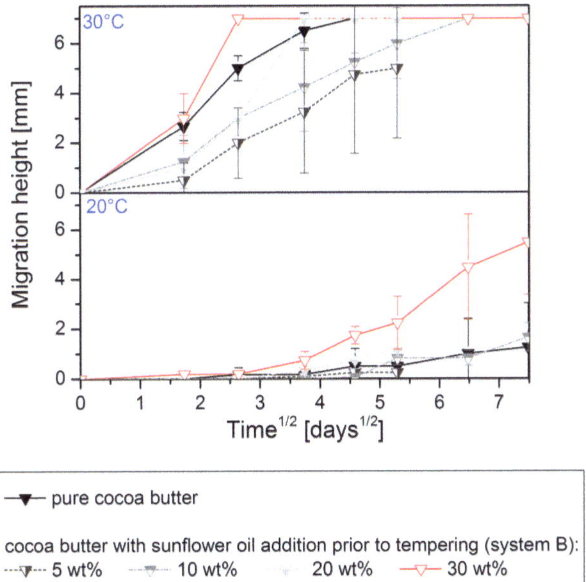

FIGURE 6.13: Upwards migration of tempered cocoa butter samples with addition of oil at different concentrations. Storage of samples at 20 °C and 30 °C.

cannot describe the experimental results. Kadivar et al. (2016) determined that addition of 5 wt% of sunflower oil to cocoa butter lead to no significant change of melting behavior, whereas 25 wt% oil resulted in a decrease of solid fat content. But they found out that 5 wt% oil addition decelerates the post-crystallization process, which leads to a significant increase in onset melting temperature and hardness of pure cocoa butter after 4 weeks of storage. It could be that some addition of oil facilitates better and more homogeneous crystallization and thus a denser fat network leading to higher resistance. Campos et al. (2010) describes that cocoa butter can crystallize into a more stable structure with addition of low melting fats. The reason is a higher mobility because of an increase in liquid fat content and thus better rearrangement of molecules within the crystal network[9]. An analysis of the spatial distribution of crystal structure and solid fat content of the different samples with oil addition could give further insight into the effect of oil addition on cocoa butter structure. This could verify the findings of this experiment.

In contrast, migration is accelerated at 30 wt% oil addition to the cocoa butter (Figure 6.12, 6.13). Differences are more pronounced for downwards migration. For very high oil addition, this might be due to formation of channels throughout the sample (Figure 6.14). Adam-Berret et al. (2011) found that tortuosity of samples with low solid fat contents are lower than for higher solid fat contents. Furthermore, it could be that a high excess amount of liquid fat lead to gravity effects. This could be tested by storage of samples with addition of 30 wt% upside down, which has not been done so far. The microstructure of the samples was analyzed to better understand the influence of oil addition on migration rates. The inner structure of cocoa butter samples with addition of oil was visualized with light microscopy and visual observation (Figure 6.14). The cross section of samples with addition of 10 and 30 wt% of oil are characterized by regular straight lines which are due to the cutting process and by less regular structures. These structural elements are distributed throughout the sample and have a length up to millimeters (Figure 6.14). The structures are more pronounced at addition of 30 wt% of oil than at only 10 wt%. These could be channels through which migration progresses.

To better understand the retarding effect of addition of a small proportion of oil in pure cocoa butter on lipid migration, migration rates are calculated for cocoa butter with oil samples (Figure 6.16). Therefore, migration height per time is approximated with a 4th order polynomial function (Figure 6.15). Migration rates are determined by taking the derivative of the migration height in dependence of time. Figure 6.16 clearly shows that migration rates with addition of 5 up to 20 wt% of oil to tempered cocoa butter show an

---

[9]This could also explain the deviation in case of 10 wt% addition in Figure 6.7. In order to enhance structure special care has to be taken during mixing of the oil to the precrystallized cocoa butter which might have been insufficient for Figure 6.7. This resulted in high standard deviations.

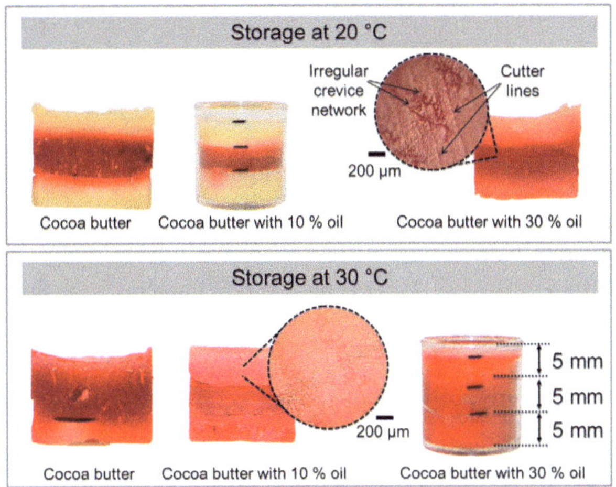

FIGURE 6.14: Photographs and microscopy images of cross sections and side views of tempered cocoa butter samples with addition of oil at different concentrations. Storage of samples at 20 °C and 30 °C.

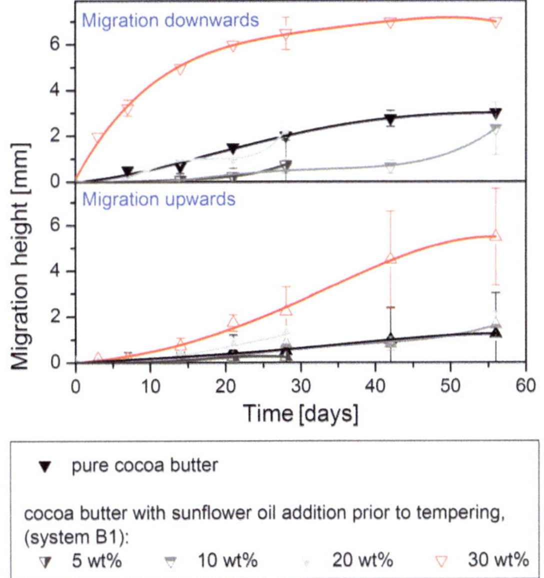

FIGURE 6.15: Migration height of cocoa butter with oil and approximated polynomials (4th order polynomial fitting used).

induction time with no upwards nor downwards migration for 3 to 7 days which cannot be seen for any of the pure cocoa butter samples. Afterwards, upwards migration rates increase for 5 wt% addition to 0.02 mm/days after 18 days of storage, with 10 wt% of oil to 0.04 mm/days after 20 days, for 20 wt% oil addition to 0.07 mm/days after maximum observation time of 28 days and for 30 wt% oil addition up to 0.15 mm/days. Subsequently, upwards migration rates approach a constant value or drop for cocoa butter samples with addition of 5, 10, 20 or 30 wt% oil addition. Downwards migration rates fluctuate more than the ones for upwards migration (Figure 6.16). After the induction time migration rates increase for samples with addition of 10 and 20 wt% of oil at 15 days and at 9 days of storage, respectively. Downwards migration rates of cocoa butter with 5 wt% of oil increase until end of observation period (28 days)[10].

6.5 Analysis of the influence of particle addition on migration through cocoa butter

Sucrose particles were added to cocoa butter resulting in 30 wt% of particles embedded in cocoa butter. To investigate the influence of particle size small icing sugar (particle size from $d_{10,3} = 8$ µm to $d_{90,3} = 79$ µm) and large household sucrose particles ($d_{10,3} = 256$ µm to $d_{90,3} = 468$ µm) were used. In addition, a suspension of 2/3 by weight sucrose refined in cocoa butter and lecithin with a final particle size between $d_{10,3} = 2.48$ µm and $d_{90,3} = 39.87$ µm, thus a sample similar to that of conventional chocolate, was investigated. Microscopy images of cross sections (Figure 6.17) show that sucrose sedimented to the bottom of each layer. Thus, the sucrose distribution was not homogeneous within the sample. Due to that, results with small and large sucrose particles have to be seen with caution. The sedimentation of large sucrose particles might lead to a barrier to migration.

Figure 6.18 shows that addition of sucrose particles at 20 °C has no effect on the migration height in upwards direction, which is constantly low with a maximum migration height of 1.5 mm after 8 weeks of storage. In general, downwards migration is more pronounced than upwards which is consistent with the previous macroscopic observations (e.g. Figure 6.5). Downwards migration height increases more rapidly for pure cocoa butter than for cocoa butter with particles after 2 weeks. Migration height of samples with sucrose particles only increase after 3 weeks but with a faster migration rate than pure cocoa butter. Thus, migration height after 6 weeks for cocoa butter with small sucrose particle is 2.25 mm and for pure cocoa butter it is 2.75 mm on average.

---

[10]Samples from sub-study presented in Figure 6.7 with 10 wt% of oil addition show a similar behavior of migration rates as untempered samples. Migration rates first decrease for 7 days (upwards) to 29 days (downwards) with a subsequent decrease. The increase in migration rate is later than that of untempered samples in case of downwards migration but earlier and less pronounced for upwards migration. This indicates that the sample might not have been crystallized as homogeneously as samples in this sub-study from Figure 6.16. This explains the high standard deviation and is a possible explanation that the 10 wt% are not consistent with this sub-study.

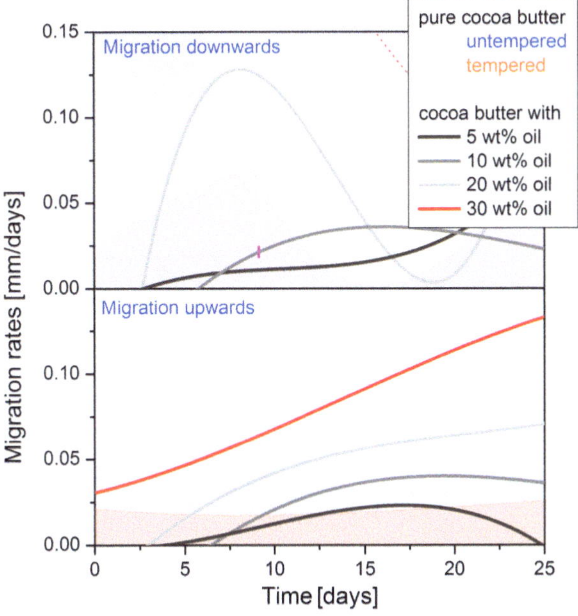

FIGURE 6.16: Migration rates in cocoa butter with oil from the derivation of the approximated migration curves.

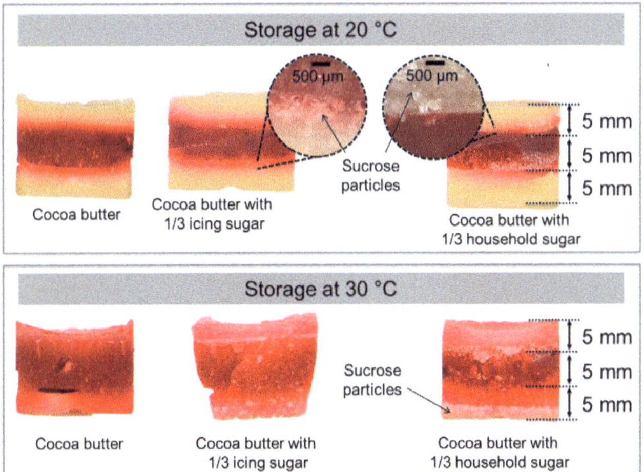

FIGURE 6.17: Photographs and microscopy images of cocoa butter samples and cocoa butter with addition of particles. Storage of samples at 20 °C and 30 °C.

Addition of small sucrose particles slightly decreases migration at 30 °C between 1 and 4 weeks of storage. With large sucrose migration heights at 30 °C are equal to that of pure cocoa butter.

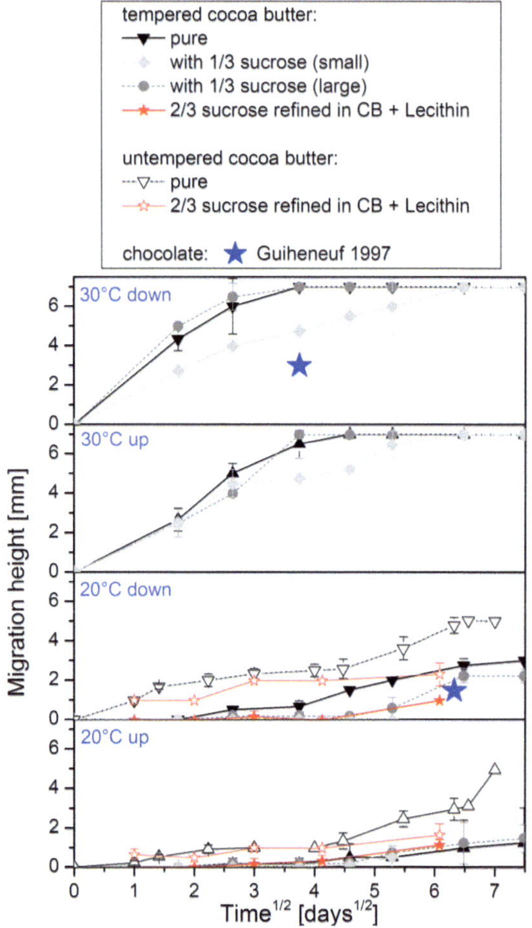

FIGURE 6.18: Migration of cocoa butter samples with addition of particles. Storage of
samples at 20 °C and 30 °C.

The cocoa butter sample with 2/3 by weight refined sucrose probably shows best how
particle addition in conventional chocolate influences migration. Typical chocolates have
a fat content of about 25 to 30 % (Wohlmuth, 2009), which is the same as the refined
sucrose particle suspension. The sedimentation effect is probably much less significant
due to higher particle content. Thereby, migration heights are about 1 mm lower after
3 weeks up to end of observation time of about 5 weeks at 20 °C for tempered as well
as untempered samples. The migration heights are in the range of that of migration
of a hazelnut filling into dark chocolate (Guiheneuf et al., 1997) at 20 °C. Migration
into refined sucrose suspension was not measured at 30 °C. But as it can be seen in
Figure 6.18, migration heights are lower for the refined sucrose suspension than for pure
cocoa butter. Thus, it is expected to be the same at 30 °C storage temperature and

would be therefore consistent with Guiheneuf et al. (1997) migration at 28 °C, which is lower than that of pure cocoa butter.

Svanberg et al. (2011c) found a pronounced effect of addition of sugar particles on microstructure of non-seeded cocoa butter with more stable crystals because of particle addition. A possible reason could be additional nucleation sites at the particle surface. However, microscope images of Svanberg et al. (2011a) did not verify that crystallization evolved from the particles. They did not observe differences due to particle addition on a local microscale but macroscopically addition of particles retarded migration (Svanberg et al., 2011c) which corresponds to Figure 6.18. Addition of small sucrose particles and sucrose refined with lecithin in cocoa butter retarded macroscopic migration. This effect was not found for large sucrose particles which is probably due to significant segregation effects. An increase in tortuosity is often proposed as a reason for reduced macroscopic migration (Svanberg et al., 2011c). However, another reason could be that formation of microstructure of the fat phase is altered due to addition of small sucrose particles[11]. The results are strengthened with the findings from Afoakwa et al. (2008b). They investigated crystal structure of dark chocolates with varying particle size distribution and composition. Thereby, the degree of crystallinity and crystal size distribution increased with increasing fat content between 25 % and 35 % and decreased with lecithin content but onset and peak temperature of melting was not altered. Thus, lower fat content chocolates need longer time to melt. Furthermore, high fat content chocolates had larger fat crystal sizes. This could be because of lower inter-particle interaction and increased free-moving plastic flow because of decreasing effect of fat being trapped in voids between particles (Afoakwa et al., 2008b). However, variation of particle size between D_{90} of 18 to 50 µm had no major impact on the crystallinity and melting properties of chocolate samples (Afoakwa et al., 2008b). Dark chocolates with larger particle sizes completely melted at slightly lower temperatures than with smaller particle sizes. This could be due to higher inter-particle aggregation and flocculation and thus it might be that the particle network provides resistance to melting[12]. However, melting enthalpy was similar with all particle sizes.

---

[11]Migration of a highly oleic filling into white chocolate is faster than into pure cocoa butter (system A, Figure 6.4). However, white chocolate contains additional ingredients. Besides cocoa butter and sucrose milk powder and milk fat is present in white chocolate. Additional, an emulsifier which probably has an effect on migration is added to the chocolate recipe to improve rheology of the chocolate. However, migration in a suspension out of sucrose refined in cocoa butter with lecithin (similar process to real chocolate manufacturing) is slightly retarded compared to pure cocoa butter (Figure 6.18). Thus, emulsifier in typical concentrations as present in chocolate cannot be a major factor for higher migration rates of white chocolate. Therefore, it is very likely that the milk component is leading to faster migration in white chocolate compared to pure cocoa butter. Milk fat is a lower melting fat and is known to softened the cocoa butter fat structure.

[12]The results in Figure 5.6 on page 79 in Chapter 5 suggest that small sucrose particles stabilize the sample structure.

6.6   Analysis of the structure dependence of migration

Previous experimental results presented in this thesis suggest that the structure of the fat phase has the highest impact on macroscopic migration at constant storage conditions among all parameters investigated. Regarding fat phase structure it was found that migration in untempered samples is faster than in tempered samples (Figure 6.5). For the experimental results of system C presented so far, the composition of the tempered and untempered samples was kept constant for all three layers. Consequently, no information can be extracted from those data on the impact of the stained layer from which migration starts compared to that of the layer into which it occurs. Thus, another experiment was conducted with varying layer properties to better understand why tempering has such a strong impact on macroscopic migration. Therefore, migration in samples with layers with alternating tempering degree (tempered versus untempered) are investigated. One batch of samples were prepared with the middle layer untempered and top and bottom layer tempered (system D1) and one batch had a tempered middle layer surrounded with untempered layers (system D2). Different layers were stained with Sudan red color in order to investigate up- as well as downwards migration from and into different layers. In general, migration into a tempered layer is slower than migration into an untempered layer (Figure 6.19 and 6.20) independent of the structure of the stained layer from which the investigated migration starts. No significant difference can be seen for migration from an untempered into a tempered and from a tempered into a tempered layer. Same holds for migration into an untempered layer. Upwards migration shows the same characteristics as downwards migration but at a lower rate. As seen before (e.g. Figure 6.5), upwards migration into the layer which has been molded on top of the stained layer is less pronounced than downwards migration into the bottom layer, which has been molded prior to the stained layer (Figure 6.20). However, the trends are the same for up- and downwards direction: Migration into tempered layers from either tempered and untempered is similar and slower than migration into an untempered layer both from tempered and from untempered layers, which again are similar. Visual observation of the interfacial region of two adjacent layers (Figure 6.21) expose that untempered phases are more coarse and less homogeneous than tempered phases independent of the adjacent layer.

The migration height per time was approximated in order to reduce noise caused by reading inaccuracy and to calculate migration rates from the experimental values. Measured values were approximated with a 4th polynomial function (Figure 6.22, Table 8.3 on page 138 in the Appendix). The approximated functions describe the experimental data and can follow the trends (Figure 6.22). The maximum deviation to any data point within all samples is less than 0.5 mm which is close to the detection limit of the visual

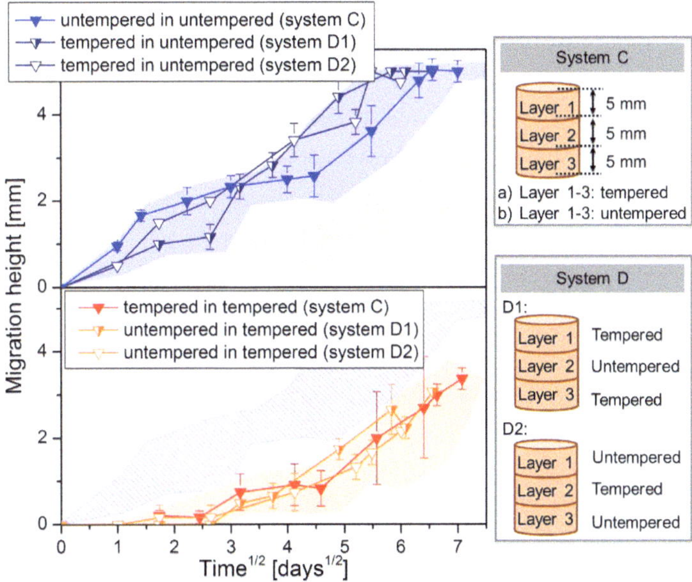

FIGURE 6.19: Downwards migration of cocoa butter samples from and into differently tempered phases. Storage of samples at 20 °C.

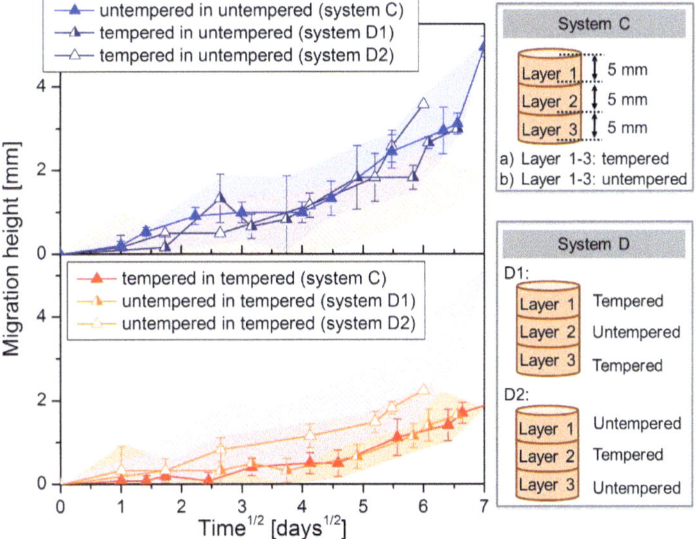

FIGURE 6.20: Upwards migration of cocoa butter samples from and into differently tempered phases. Storage of samples at 20 °C.

observation. Mean deviation is less than 0.3 mm for all approximations and in most cases less than 0.2 mm. The maximum sample height was 5 mm. Thus, migration rates were only considered until a height of 4 mm is reached to avoid boundary effects. More than 4 mm migration height was observed for downwards migration of untempered into untempered (system C) cocoa butter layers, for downwards migration of tempered into untempered cocoa butter layers of system D1 and D2. The approximated curves were used to get migration rates for tempered and untempered cocoa butter samples per observation time (Figure 6.23) and as ranges over the entire observation time (Figure 6.24).

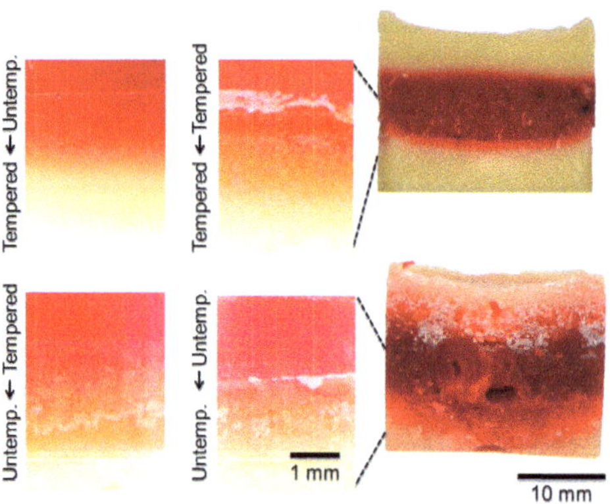

FIGURE 6.21: Photographs of cocoa butter samples with migration from and into differently tempered phases. Storage of samples at 20 °C.

Migration rates of tempered cocoa butter layers into tempered cocoa butter is below 0.1 mm/day for upwards and downwards migration over the entire observation time period (Figure 6.23). In contrast to that, the migration rates from untempered into untempered cocoa butter layers is in general higher and less constant over the observation time. Downwards migration rates of these samples are very high and decrease for the first 2 weeks until a migration rate of 0.002 mm/day at 15 days of storage followed by an increase until the maximum sample height was reached. The change in migration rate is not that pronounced for upwards migration. However, upwards migration rates into untempered layers also drop within the first three weeks. Downwards migration rates into untempered layers from untempered (system C) and from the tempered layer of system D2 are very similar. Downwards migration rates from and into tempered layers of system C follows a similar trend but at a much lower amplitude. Downwards migration of both samples of system D1 (from untempered into tempered as well as from tempered into untempered) seems to be opposed to that of the D2 and C systems. Migration rates first slightly increase until a maximum after 1 to 2 weeks with a subsequent decrease with a local minimum after about 5 weeks and another increase. However, as Figure 6.22 shows, the migration height development of these samples is similar. Additionally, it

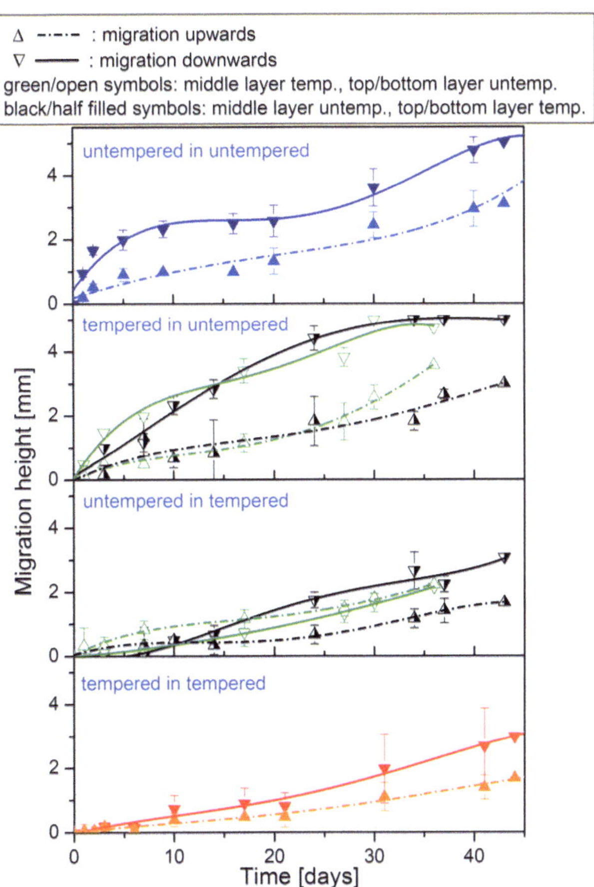

FIGURE 6.22: Migration height of cocoa butter and approximated polynomials (4th polynomial fitting used).

has to be noted that slight difference in the fit functions of migration height over time already leads to significant differences in rates.

Migration rates into upwards direction are for all systems similar. Only difference can be seen at the end of the observation period after 4 weeks with increasing rate for upwards migration into untempered samples (mostly system C and D2, system D1 rather constant). In contrast to that, upwards migration rates into tempered layers do not increase at the end of observation time. Figure 6.24 also illustrates that the migration rates of untempered samples are less constant than the ones of tempered samples.

A possible reason for a decrease in migration rates within the first days of storage is post-crystallization of cocoa butter. Rousseau and Sonwai (2008) measured an increase in solid fat content of tempered pure cocoa butter from 67 % to 70 % and for tempered

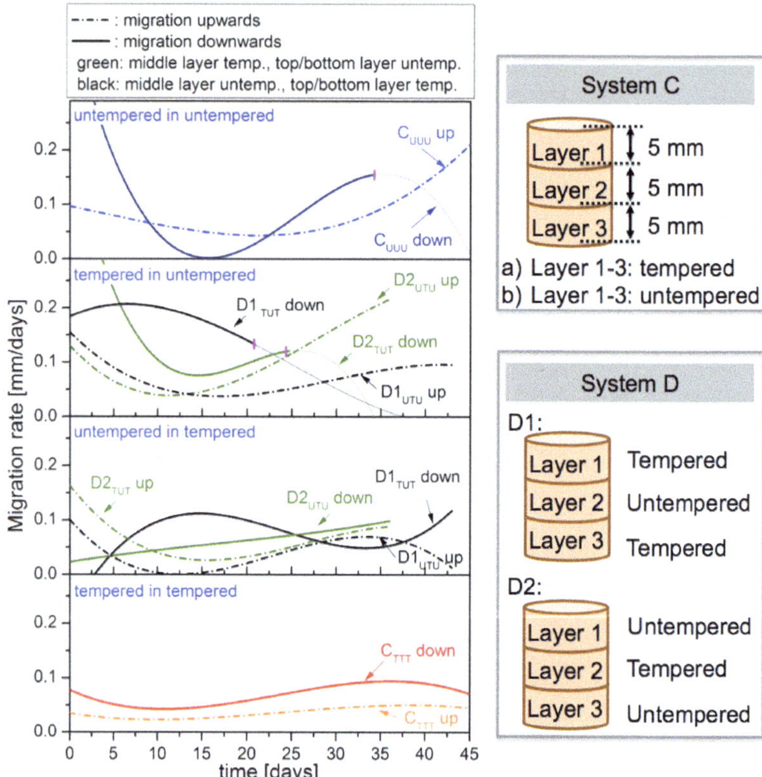

FIGURE 6.23: Migration rates in cocoa butter from the derivation of the approximated migration curves.

chocolate from 81 % to 84 % possibly due to post-crystallization. Thus, this effect is more pronounced in case of untempered samples which explains the fast decrease in migration rates from high values (due to possibly low solid fat content) to low values of migration rates for untempered samples. It might be that the structure changes greatly for untempered cocoa butter which leads to notable change in migration rate. Migration rates of tempered samples are more stable probably due to little change in microstructure. Tempering aims to establish a stable and homogeneous crystal structure. Consequently, the overall structure of the sample is more stable in case of tempered samples compared to untempered samples which crystallized in an uncontrolled manner. As a result untempered samples contain less stable crystals which rearrange with storage time leading to structural changes.

In general, smaller migration rates mean either that the driving force is lower or the resistance against migration is higher (Ziegler, 2007). In case of diffusion, a gradient in triglyceride content is a possible driving force for migration in cocoa butter according

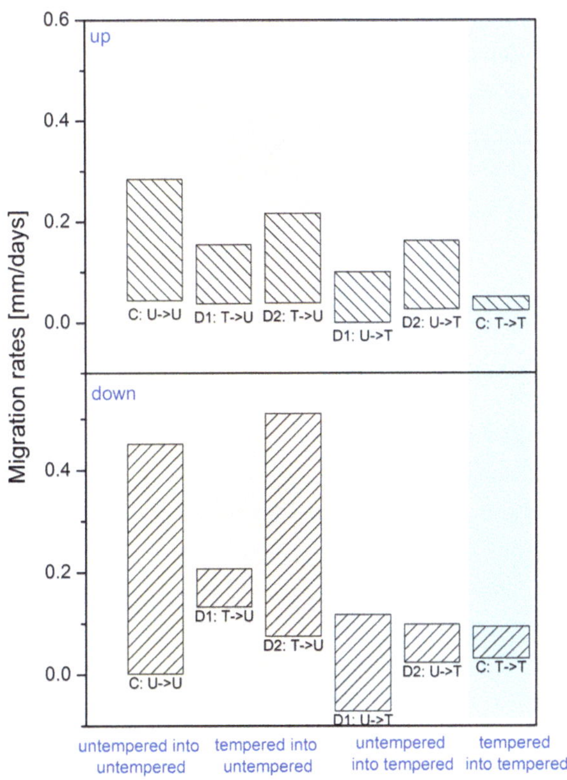

FIGURE 6.24: Migration rate ranges from minimum to maximum rates over observation
time. U = untempered layer; T = tempered layer

to Ziegleder et al. (1996a). The difference in composition of the mobile phase and gradient of liquid fat between both layers is expected to be higher in case of two different
polymorphic phases. Narine and Marangoni (1999a) for example measured a lower solid
fat content for untempered cocoa butter at temperatures between 20 and 35 °C than
tempered samples measured by Torbica et al. (2006) (Figure 2.5 on page 16). However, no increase in migration rate is observed when the difference between both layers
is enhanced due to variation in precrystallization grade of adjacent layers. Migration
rates do not significantly differ between two untempered compared to a composite of an
untempered and tempered layer (Figure 6.20). Consequently, it seems to only matter,
at least under the investigated conditions, into which layer migration takes place. This
possibly is a result of the impact of structure, which is greater than that of a gradient

in solid fat content and also triglyceride content. [13] Consequently, reduced migration rates might be a result of an increase in resistance due to a denser structure. Thus, in case of diffusion induced transport this would mean that resistance, which is higher in case of tempered layers than in case of untempered ones, is the main factor controlling migration rate. Another explanation is that the transport is driven by capillary action. Transport through capillaries is slower through small capillaries than larger ones (see for example equation 2.6 on page 27). This is strengthened by the findings of Figure 6.5 and the related discussion regarding stepwise migration.

6.7 Prediction of migration based on theoretical models

To better understand the experimental findings, lipid transport was modeled through porous glass as a simplified matrix system. Subsequently, the model was applied to migration through cocoa butter. The aim of the simulations is to evaluate the underlying mechanism of the observed macroscopic migration discussed above. Therefore, flow of mobile lipids through nanometer large channels was calculated based on Darcy's equation with capillary pressure as the main driving force. In general, both diffusion and capillary rise are widely discussed transport mechanisms for lipid migration in chocolate. And most probably both mechanisms play a role (e.g. Aguilera et al., 2004, Rousseau and Smith, 2008). Diffusion based migration in chocolate has been extensively modeled before (e.g. Galdámez et al., 2009, Khan and Rousseau, 2006, Ziegleder et al., 1996a,b,b, Ziegleder and Schwingshandl, 1998, Ziegler, 2007). However, migration in chocolate cannot be solely described with a simple Fickian diffusion equation (Choi et al., 2007, Green and Rousseau, 2014). Thus, this simulation based on capillary rise was conducted to evaluate whether migration in chocolate can be described with the equations of viscous flow induced by capillary pressure. The experimental findings from Figure 6.4 suggest that migration happens through confined spaces with porosity in the nanometer range. Migration of hydrocarbons through the porous glass has been successfully modeled based on Darcy's equation (Equation 2.6 on page 27) and capillary pressure as main driving force by Grüner et al. (2016). Furthermore, the experimental results presented in the previous parts showed that structure through which migration takes place has a major impact on transport rates. This is also true for capillary rise.

---

[13]Migration of only dye might also occur as discussed in Section 6.1 on page 90. Thus, because the concentration gradient of the color is kept constant, the migration would also in case of diffusion of only the dye only depend on the structure of the phase it diffuses into.

6.7.1 Simulated migration heights through porous glass

The flow profile in a capillary does not necessarily develop within the initial cross sectional area with the radius r_0 (Figure 3.8 in Chapter 3 on page 53). The hydrodynamic radius r_h can be larger than the initial radius r_0 due to e.g. slippage or it can be smaller due to e.g. sticking (Grüner et al., 2016). Calculating the ratio f_h (Equation 3.10, page 53) from the slope of migration height and square root of time (Figure 6.4) with Equation 2.11 on page 29 and Equation 2.13 on page 29, gives a value of 0.86. A value of $f_h = 0.86$ corresponds to a hydrodynamic radius of $r_h = 2.9$ nm under the assumption that a sticking of a monolayer of water at the wall reduces the radius at which the meniscus forms to $r_l = r_0 - 0.25$ nm (Figure 3.8 illustrates the model). Figure 6.25 shows the imbibition of oil into porous glass for different ratios f_h of hydrodynamic and initial pore radius. The simulation was based on data for the porous glass from Grüner et al. (2016) (initial pore radius $r_0 = 3.4$ nm, tortuosity $\tau = 3.6$ and volume porosity $\phi_0 = 0.3$ as well as $\phi_i = 0.245$) and on the macroscopic properties of sunflower oil as mobile phase (surface tension $\gamma = 35$ mN/m[14], viscosity $\eta = 60$ mPa s, density $\rho = 925$ kg/m^3) and the measured apparent contact angle $\theta = 19.3°$ as well as an initial time of $t_0 = 0.9$ days, which is based on the experimental results from Figure 6.4. From the simulation in Figure 6.25 it can be seen that the experimental migration height is in the range of $f_h = 0.7$ to 0.9 corresponding to hydrodynamic radii of $r_h = 2.38$ nm to 3.06 nm. Simulated migration rates are less for smaller capillary radii and larger for higher ones, respectively.

Grüner et al. (2016) found that the hydrodynamic capillary radius r_h is smaller than the initial pore sizes r_0 during imbibition through the porous glass. The modal values of the pore size distribution from migration measurements were $\bar{r}_h = 2.77$ nm (compared to an initial value before migration of $\bar{r}_0 = 3.40$ nm). They explain the reduction of $\delta = \bar{r}_0 - \bar{r}_h = 0.63$ nm with sticking of a monolayer of water (0.25 nm) and a hydrocarbon molecule (~ 0.35 nm which is equal to the diameter of the hydrocarbon chain) at the pore wall (Grüner et al., 2016). Thus, the reduction from $\bar{r}_0 = 3.40$ nm to $\bar{r}_h \approx 2.9$ nm on average might be a result of sticking of a monolayer of water which would be roughly ~ 0.25 nm according to Grüner et al. (2016) and a triglyceride molecule. The difference is in the range of short spacings (on average 0.4 nm) of lipids in the crystalline lattice as determined by small angle X-ray scattering (Wille and Lutton, 1966). The diameter of a triglyceride molecule was estimated with calculation of the distance between two molecules in a molecular dynamic environment (Figure 6.26). Figure 6.26 shows a POP triglyceride molecule in vacuum and the difference between molecules in radial direction at the center of the molecule is between 0.26 and 0.55 nm.

---

[14]Surface tension of sunflower oil is $\gamma_{Oil} = 32.7 \pm 0.1$ mN/m and does not significantly change due to addition of 1 wt% Sudan red, which is $\gamma_{Oil+1wt\%dye} = 32.9 \pm 0.2$ mN/m at a temperature of T = 22 °C.

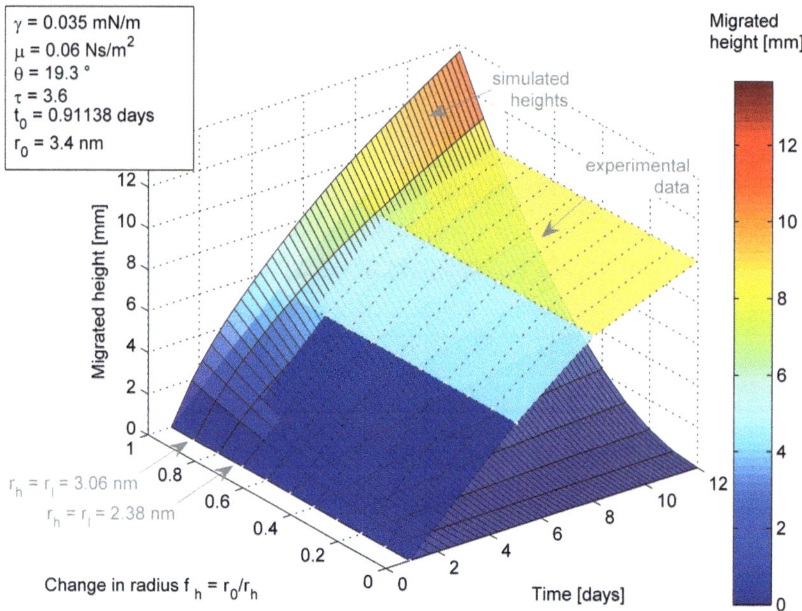

FIGURE 6.25: Simulated migration height of oil through a glass cylinder with a tortuosity of 3.6 based on capillary rise. Variation of ratio between initial pore radius and hydrodynamic radius.

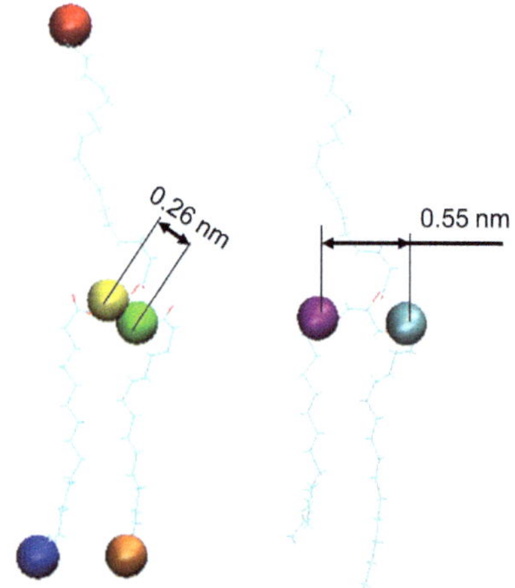

FIGURE 6.26: POP triglyceride molecule in vacuum from molecular dynamic simulation.

Grüner et al. (2016) found that squalene, which is a branched hydrocarbon, lead to an additional reduction of 0.3 nm. A triglyceride molecule has three fatty acid chains and is thus expected to be larger than a single hydrocarbon chain and the squalene molecule. The determined hydrodynamic radius r_h for hydrocarbon chains imbibing into the porous material is slightly smaller than during imbibition of lipids (Figure 6.25) (Grüner et al., 2016). The formation of the flow profile in confined spaces is complex and can involve for example slippage in addition to sticking, which alter the cross section of the flow profile. Furthermore, the migrant is a fat mixture and it might be that specific components preferentially stick at the wall. In addition to that, it has to be noted that separation might occur leading to part of the fat mixture imbibing at a faster or slower rate because of e.g. formation of different contact angles. Assuming that the radius, at which the meniscus forms r_l, is equal to the hydrodynamic radius r_h, leads to a simulated hydrodynamic radius of 2.3 nm which is even smaller than the one presented in Figure 6.25 and that for hydrocarbon chains by Grüner et al. (2016). The remaining cross section where the flow profile develops would be $A_{cap} \approx 26.4$ nm^2, which would be ~ 275 times the cross sectional area of a hydrocarbon molecules with a diameter of 0.35 nm (Grüner et al., 2016) ($A_{mol} \approx 0.10$ nm^2) and with regard to packing would lead to $n \approx 203$ triglyceride molecules in that cross section (Equation 3.11 on page 53)[15]. These dimensions correlate to values of Kn = 1.15 nm/2.9 nm = 0.38 (Equation 2.7 on page 28 with a mean path of 1.15 nm which is based on a density of 925 kg/m^3 and a molecular weight of 833.4 g/mol for a POP triglyceride molecules) and is thus beyond that of continuous flow with Knudsen numbers of smaller than 0.01. According to Ziarani and Aguilera (2012) this region is characterized by two possible migration mechanisms, slip (continuum) or diffusion. It is a region where traditional fluid dynamics might not be capable to describe the flow anymore. However, Ziarani and Aguilera (2012) also conclude that conventional equations could be applied even if still questionable and a Knudsen's correction factor is recommended.

In addition, it is to be noted that standard deviations due to accuracy of experimental set-up is at least 0.25 mm and fluid properties inside the confined spaces cannot be directly measured and thus macroscopic values were used. Small deviations in properties lead to changes in simulated migration heights. Furthermore, the porous structure is characterized by a pore size distribution rather than a single capillary radius. The effect of contact angle, viscosity and tortuosity are varied to analyze their impact on migration height. Migration based on capillary rise is significantly influenced by the contact angle with decreasing migration rates at higher contact angles (Figure 6.27). There is no migration at contact angles larger than $\theta = 90\,^\circ$ (Figure 6.27). Migration rates increase more rapidly with an increase in contact angle at high angles, whereas

---

[15]Under the assumption that the radius, at which the meniscus forms r_l, is equal to the hydrodynamic radius r_h, the simulated hydrodynamic radius is with 2.3 nm even smaller than the one presented in Figure 6.25. This leads to a reduction of initial radius r_0 of an additional 0.88 nm to the 0.25 nm of the water layer. This would be about 2.5 times of the diameter of a hydrocarbon (Grüner et al., 2016) and would lead to only $n \approx 25$ triglyceride molecules in that cross section (Equation 3.11).

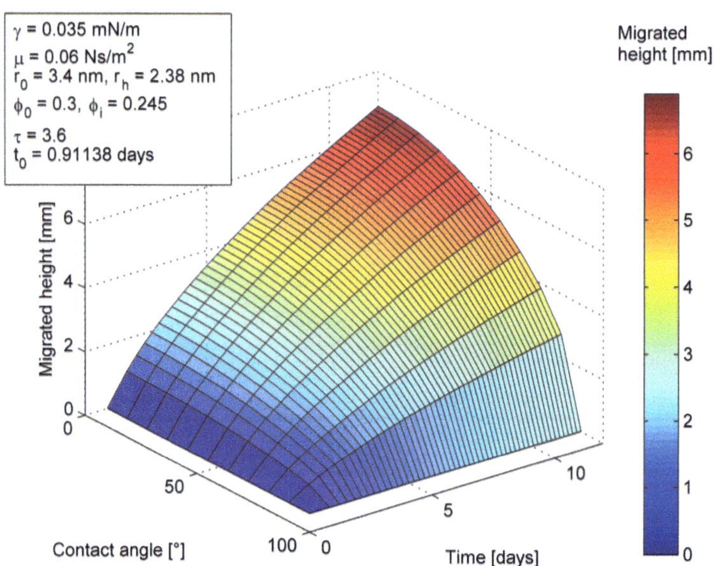

FIGURE 6.27: Simulated migration height based on capillary rise. Variation of contact angle.

no significant difference can be seen at contact angles of about $\theta = 30\,^\circ$ to complete spreading ($\theta = 0\,^\circ$). Capillary rise is slowed down at large contact angles because of a repulsive effect of liquid and capillary wall. Besides contact angle, viscosity mainly impacts transport through capillaries (Figure 6.28). Migration at lower viscosities is significantly faster than at values in the range of liquid oils (Bürkle, 2011) (Figure 6.28). Viscosity has its highest impact on migration height at values of less than $\eta = 0.2$ Ns/m² compared to higher viscosities up to $\eta = 1$ Ns/m². In addition, migration is impacted by the tortuosity which can be seen from Figure 6.29. Thereby, migration heights are higher for low tortuosities.

6.7.2   Simulated migration heights through cocoa butter

As the measured migration heights through cocoa butter and white chocolate are within the range of porous glass with pore sizes of about 3.5 nm (Figure 6.4), gravity can be neglected. For example, a capillary radius of 3.5 nm and the given parameters ($\eta = 60 \cdot 10^{-3}\ Ns/m^2$, $\sigma = 35 \cdot 10^{-3}\ N/m$, $\rho = 925\ kg/m^3$, g $= 9.81\ m/s^2$, $\phi = 45°$, $\theta = 19.3°$, Equation 2.9) gives an equilibrium height of 3 km, which is much larger than the maximum sample height. Therefore, migration rates through tempered and untempered cocoa butter is modeled with Darcy's equation with the Laplace pressure as the driving force based on the explanation given in Grüner et al. (2009) neglecting

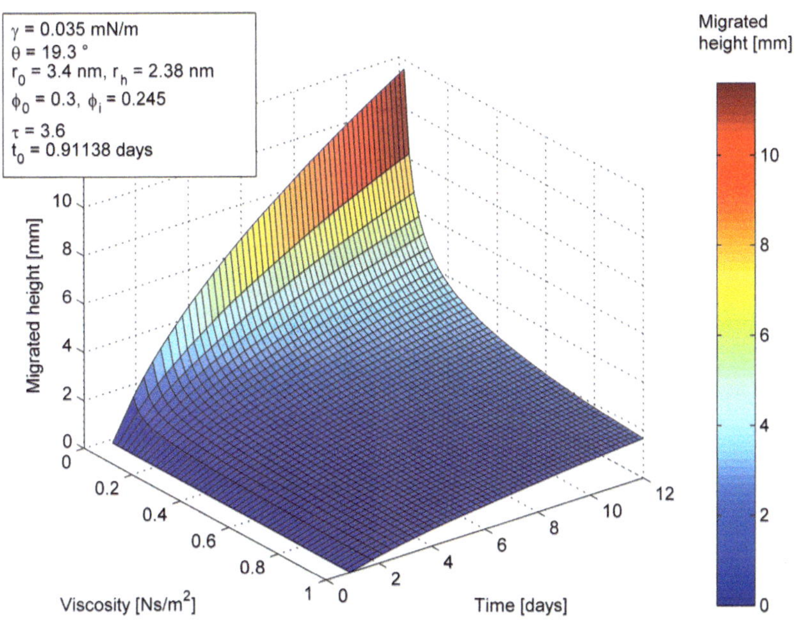

FIGURE 6.28: Simulated migration height based on capillary rise. Variation of viscosity.

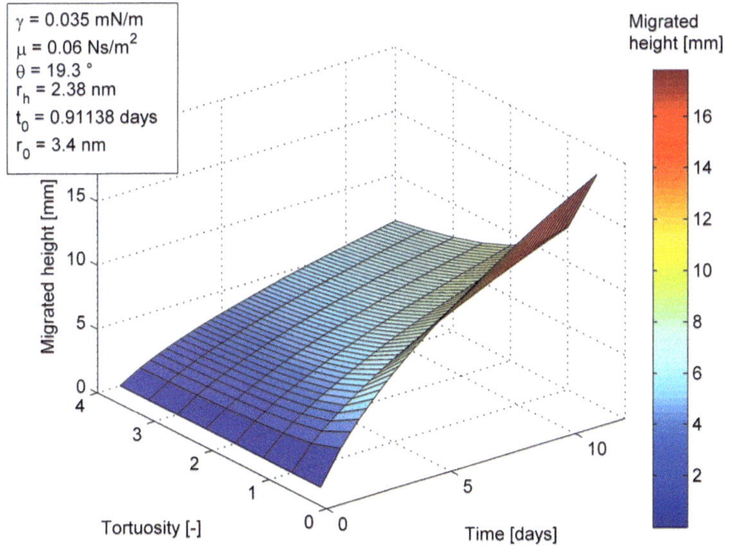

FIGURE 6.29: Simulated migration height based on capillary rise. Variation of tortuosity.

gravity. The initial tortuosity τ was set to 2 which is in the range of 2.3 (at 75 % solid fat content) to 1.3 (at 15 % solid fat content) measured by Adam-Berret et al.

(2011). The experimental data for tempered and untempered cocoa butter are fitted to Equation 3.7 (page 52) (Figure 6.30 and Figure 6.31). As seen and discussed e.g. in Figure 6.5, migration rates are not constant over the time, thus fits were carried out for experimental data points over different time periods.

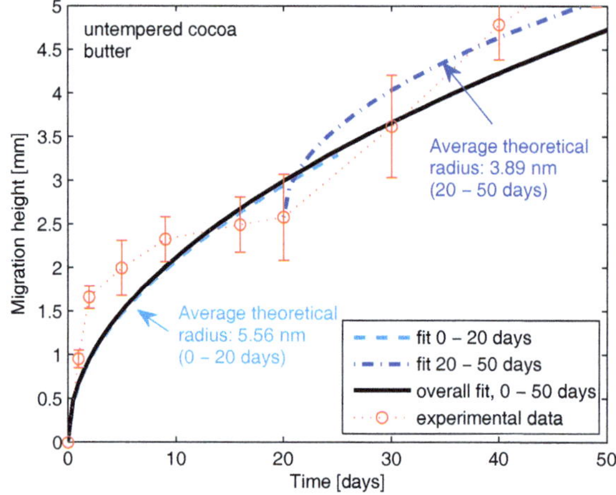

FIGURE 6.30: Fitted migration based on Equation 3.7 on page 52 and experimental data for untempered cocoa butter (Figure 6.5).

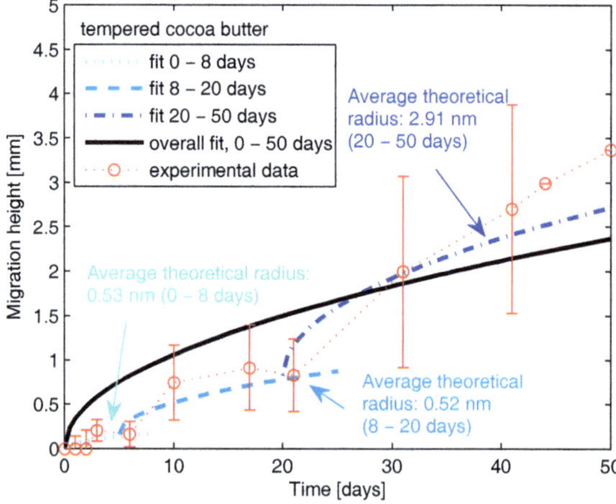

FIGURE 6.31: Fitted migration based on Equation 3.7 on page 52 and experimental data for tempered cocoa butter (Figure 6.5).

Further data for the properties of the mobile components would increase the precision of estimation of single parameters such as radius from the fitting parameter a (Equation 3.7

on page 52). More information on the surface tension, viscosity, contact angle and about the porous geometry (capillary radius and average tilt angle of capillaries or tortuosity) is necessary to finally conclude whether capillary rise is a plausible transport mechanism.

The theoretical capillary radius (Equation 3.14 on page 55) based on the simulation increases over time for tempered cocoa butter from 0.53 nm to 2.91 nm and decreased over time for untempered samples from 5.56 nm to 3.89 nm. This is consistent with average migration rates over the different time periods (Table 6.1). A decrease in migration rates, which resulted in an increase of theoretical radius, might be due to dissolution due to lipid migration as seen with small angle X-ray scattering (Chapter 4). A decrease in radius could be a result of post-crytsallization. Structure changes might also be a consequence of recrystallization of cocoa butter as discussed in Section 6.3 on page 98. Svanberg et al. (2011c) observed that the microstructure of non-seeded samples evolved with storage time and resembled the ones with seeding after one week. This could explain the findings from Figure 6.22, 6.30 and 6.31 and the fact that approximated capillary radii of tempered and untempered samples approach each other after 20 to 50 days. Directly after preparation large differences in migration rate between seeded and non-seeded samples were seen by Svanberg et al. (2011c). Non-seeded samples had significantly higher local migration rates than seeded samples. Additionally, standard deviation was higher indicating more heterogeneous structure. The difference in migration were less pronounced after 1 week of storage. The seeded samples did not change significantly and no substantial post-crystallization was detected. Local migration rates, measured with FRAP technique, were constantly low (Svanberg et al., 2011c). These findings correspond with Figure 6.23. Migration rates for untempered samples are high within the first week but decrease with a minimum after about 2 weeks (Figure 6.23). In contrast, the migration rate of tempered samples is on average constant over time, so that rates after 1 to 2 weeks for tempered and untempered samples are similar. Thus, both the simulation and the migration rates from the 4th polynomial fit curve illustrate that the migration rate over time increase for tempered and decrease in case of untempered cocoa butter samples. The simulations of migration in cocoa butter presented in Figure 6.31 and Figure 6.30 do not consider the differences between initial radius r_0, hydrodynamic radius r_h and radius at which the meniscus forms r_l depicted in Figure 3.8 in Chapter 3 on page 53. Detailed information on porosity and initial radii are needed for a deeper analysis of lipid flow through cocoa butter. In the investigated range it can be seen that the higher the initial radius r_0, the higher the hydrodynamic radius r_h at constant migration rates (Figure 6.33 and Figure 6.34). The radius is higher for high migration rates than for low ones[16]. The relationship between hydrodynamic and

---

[16]The results for larger effective diffusion coefficients and larger initial radii can be found in Figures 8.3 and 8.3 in the Appendix on page 139.

initial radius is non-linear and asymptotic, which results in an increase in the difference between both radii $\Delta r = r_0 - r_h$.

TABLE 6.1: Average slope of downwards migration height for tempered (T) and untempered (U) cocoa butter samples from system C (Figure 6.23) in comparison to the simulated rate of migration (a in Equation 3.7 on page 52) from Figures 6.30 and 6.31.

Sample: days	slope $[\text{mm/day}^{0.5}]$	slope^2 $[\text{mm}^2/\text{day}]$	D_{eff} $[\text{m}^2/\text{s}]$	simulated a $[\text{m}^2/\text{s}]$	D_{sim} $[\text{m}^2/\text{s}]$
U: 0 - 10	2.09E-01	2.18E-02	2.53E-13	2.93E-12	5.86E-12
U: 0 - 20	1.11E-01	6.18E-03	7.15E-14	2.52E-12	5.04E-12
U: 20-50	7.87E-02	3.10E-03	3.58E-14	1.23E-12	2.46E-12
T: 0 - 8	5.76E-02	1.66E-03	1.92E-14	2.25E-14	5.50E-14
T: 8 - 20	4.81E-02	1.16E-03	1.34E-14	2.22E-14	4.44E-14
T: 20-50	7.83E-02	3.06E-03	3.54E-14	6.91E-13	1.38E-12

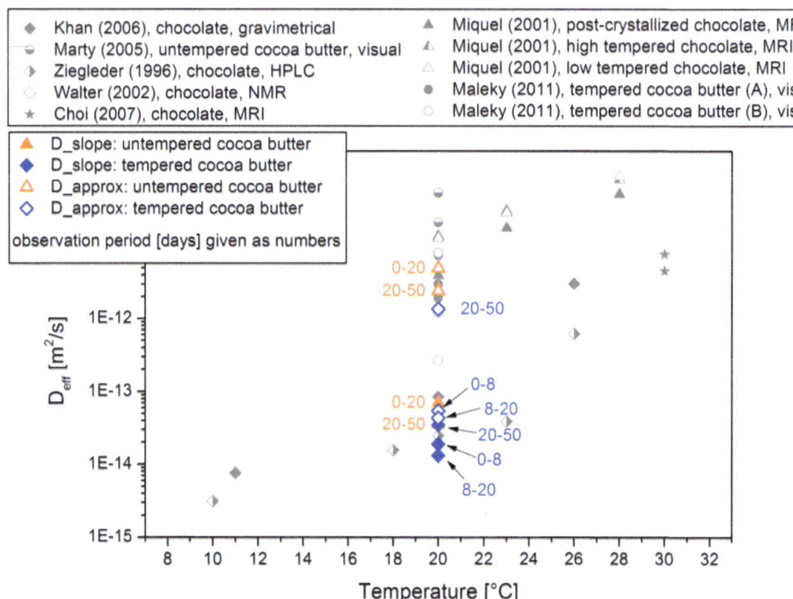

FIGURE 6.32: Calculated migration coefficients (closed symbols: D_{eff} from fitted curves of experimental data and open symbols: D_{sim} from the simulations) of tempered and untempered cocoa butter for different time spans in comparison to literature values of diffusion coefficients.

The approximated diffusion coefficients from the fit D_{approx} as well as from the migration rates D_{slope} are listed in Table 6.1 and illustrated in Figure 6.32. Khan and Rousseau (2006) determined a diffusion coefficients of $8.6 \cdot 10^{-10} cm^2/s = 8.6 \cdot 10^{-14} m^2/s$ at 20 °C and $3.1 \cdot 10^{-8} cm^2/s = 3.1 \cdot 10^{-12} m^2/s$ at 26 °C for hazelnut into dark chocolate. This is in the same range of migration from a layer of tempered into tempered cocoa butter (Table 6.1). Marty et al. (2005) calculated apparent diffusion coefficients of $5.41 \cdot 10^{-7} cm^2/s$ which is equal to $5.41 \cdot 10^{-11} m^2/s$ through their untempered cocoa butter sample.

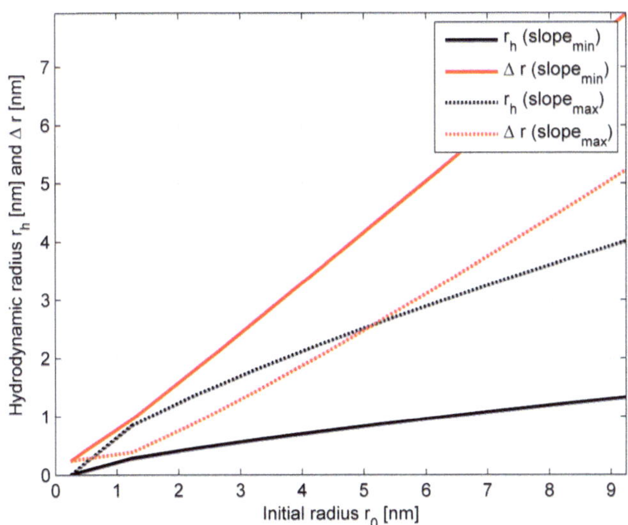

FIGURE 6.33: Change of hydrodynamic radius r_h and difference between initial and hydrodynamic radius $\Delta r = r_0 - r_h$ for different initial radii r_0 for two different effective diffusion coefficients D_{eff} with a slope of 0.1 mm/day$^{0.5}$ and 0.3 mm/day$^{0.5}$. Data for 1 mm/day$^{0.5}$ and 2.5 mm/day$^{0.5}$ are given in the Appendix.

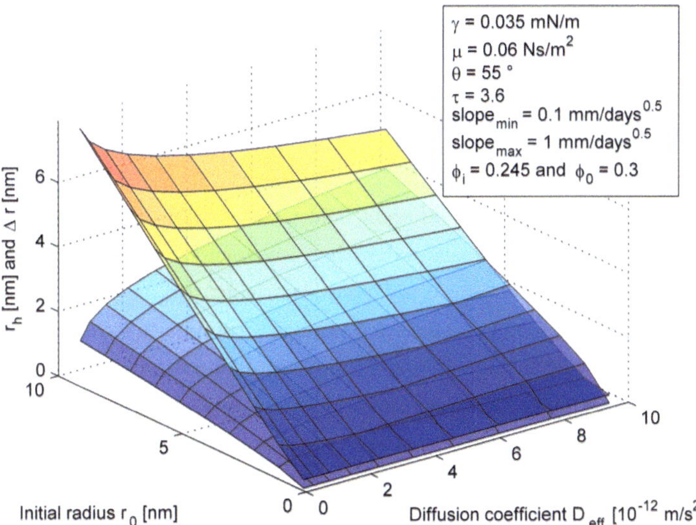

FIGURE 6.34: Change of hydrodynamic radius r_h and difference between initial and hydrodynamic radius $\Delta r = r_0 - r_h$ for different initial radii r_0 with variation of effective diffusion coefficient D_{eff} induced by different slopes in migration height per square root of time. Data for 1 mm/day$^{0.5}$ and 2.5 mm/day$^{0.5}$ are given in the Appendix.

6.8 Discussion of possible migration mechanisms based on macroscopic observation of migration

Prediction of migration pathways and the underlying transport mechanism in porous media is a challenge. Transport phenomena in porous material is probably a combination of different mechanism, which can be distinguished between free molecular flow and continuum flow, including diffusion and viscous flow as well as thermal transpiration and thermal diffusion (Bird et al., 2007).

Lipids in chocolate could migrate through the continuous matrix fat phase, through the network of particles or through cavities which can be inside the fat or particle phase or at the interface of both phases (Figure 6.35). The matrix phase of chocolate is hydrophobic whereas the particles are more hydrophilic. The remaining cocoa butter of cocoa solids are counted to the continuous fat phase. The sizes of possible pathways vary from a

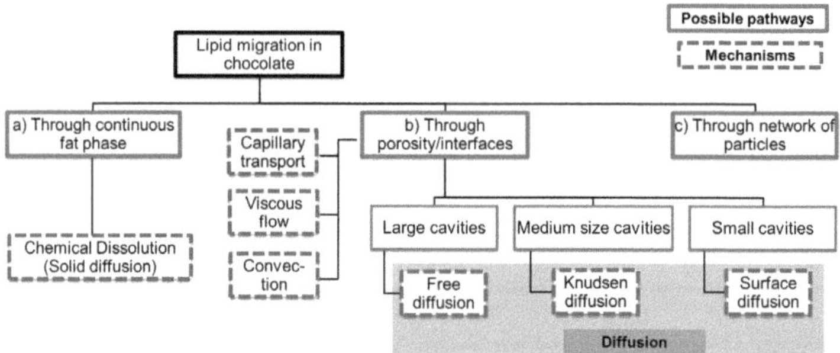

FIGURE 6.35: Overview over possible migration pathways and mechanisms.

molecular level between adjacent lipid molecules (Figure 2.2 on page 6) to microscopic pores like the ones visualized with X-ray tomography (e.g. Figure 5.7 on page 80). The migration mechanism is largely determined by the pathway and structure of the material (Figure 6.35). Migration through a continuous solid phase can appear through solid diffusion or through dissolution. Dissolution is limited to migrants which are able to dissolve the phase through which migration takes place. In case of lipid migration in chocolate this is possible for the continuous fat matrix (e.g. Figure 4.6 on page 63).

The migration mechanism in chocolate is probably due to multiple processes such as diffusion and capillary rise which take place simultaneously (Marty et al., 2005, 2009). The experimental results from this study show that migration is mainly influenced by the structure of the phase into which it takes place. Marty et al. (2005) concluded that structure may change which then impacts migration rates. This could explain the variation of migration rates with observation time.

Differences in structure of tempered and untempered samples are probably the reason for higher migration rates in case of no tempering. X-ray tomography revealed that the structure of tempered cocoa butter is more homogeneous and denser (Chapter 5) and more stable (Chapter 2, e.g. Section 2.2.2.5 on page 17). Tomography images reveal that besides crystal structure, also microstructure is highly dependent on the precrystallization step (Figure 5.3 on page 75). In general it can be seen that migration into untempered cocoa butter is faster than into tempered fat phase (Figure 6.5). This is consistent with the findings from Maleky and Marangoni (2011a), who correlated the crystalline structure to oil migration with a less permeable structure leading to slower oil migration. That would mean that the untempered samples have a looser crystalline structure. Thereby, the looser the structure and the more voids, the faster migration. Furthermore, it was found that migration is faster into layers molded prior to that from which it starts. A reason could be that the solidified surface layer of the bottom layer is partially melted when liquid cocoa butter is poured on top of the bottom layer or that the bottom layer acts as crystallization seed for the top layer. The importance of structure strengthened capillary rise as a main transport mechanism but does not exclude diffusion. In case of diffusion, a dense structure acts as resistance and thus leads to lower migration rates. In case of capillary rise, smaller channels lead to slower migration as well. The migration of the lipid filling into porous glass clearly showed that lipids can migrate through very confined spaces and pores in the size of $r_0 = 3.4$ nm. The simulations illustrated that lipid transport through the porous glass material can be described with Darcy's equation driven by capillary pressure. Migration into chocolate is in the range of that of porous glass. The transport rates of lipids through cocoa butter were calculated based on the model and can describe the migration. Applying the model to the observed migration rates leads to very small values of calculated capillary sizes. Thus, transport is probably in a transition regime between viscous flow and diffusion. Validation of the model used to describe transport through porous glass for chocolate requires a better understanding of the matrix properties, namely pore radius distribution in the relevant range, which is in the nanometer length scale, tortuosity and porosity. Furthermore, information on whether porosity in chocolate is filled with liquid oil or is an open space would help to identify relevant migration mechanism of lipids through chocolate. In addition, it was noted that migration rates changed possibly due to structure changes, which should be integrated into the model. Additionally, it has to be further investigated in how far the properties of the mobile phase, such as viscosity and surface tension, change over time due to e.g. dissolution of matrix components or recrystallization.

7

Conclusion

This thesis investigates possible migration pathways and mechanisms of lipids in crystalline fat suspensions such as chocolate. Migration of lipids in chocolate is associated with formation of fat blooming leading to consumer rejections and thus presents a major quality issue of the confectionery industry.

Chocolate is a multicomponent material consisting of particles dispersed in a fat phase, which is mainly cocoa butter. The experimental results showed that chocolate is a highly porous, densely packed material. Porosity within the chocolate is present at different length scales. X-ray tomography showed that the micrometer sized particles are closely packed within the chocolate. The gap between the solid particles is in most cases less than a micron. Exceptions are cracks and voids in the micrometer range which are present throughout the entire chocolate sample. Visual observation of all images showed that the arrangement of particles and distribution of pores and cracks seems to be independent of the position within the sample. The images show that the particles are closely packed mostly with flat surfaces densely attached against each other. Cracks with a width of some microns and a length of up to several 100 µm were revealed. A possible explanation for crack formation is the development of local stresses, which result from a difference in contraction of suspended rigid particles and the surrounding lipid matrix during solidification of the chocolate mass. The lipid matrix changes from a viscous liquid to a brittle solid material by crystallization accompanied with significant contraction, whereas the particles do not considerably contract. This is supported by lack of cracks in pure cocoa butter and increasing detection of cracks with increasing particle amount. Cracks are propagating mainly in the circumferential hoop direction which is probably due to radial stress formation due to inhomogeneous temperature gradients within the sample. It is suggested to further investigate the crack formation and propagation in multicomponent food materials with the aim to adjust the processing conditions to produce defectless microstructures. Therefore, it is proposed to compare chocolate microstructure with other materials composed of hard particles embedded in a softer ductile continuous phase such as highly filled composites or concrete. Furthermore, tomography analysis revealed that the porosity in the micrometer range is higher for untempered samples than in case of precrystallization. This effect is less pronounced at higher solid particle concentrations. Small angle X-ray scattering (SAXS) confirmed that the chocolate particles are highly porous in the nanometer range. It might be that convective flow of liquid lipid fractions through the porosity of the chocolate at different length scales plays a significant role as a transport mechanism leading to fat blooming.

Experimental results of lipid migration in chocolate and model systems presented in this thesis support the understanding of the migration process. A decrease of SAXS signal due to oil migration into the chocolate components suggest that oil rapidly migrates into pores in the nanometer range. This effect was less pronounced for samples with lower porosity such as pure cocoa butter where no migration into nanometer pores was detected within measurement time. This can be explained with a lower SAXS signal and thus less porosity than for the powders with cocoa butter. Consequently, less oil can migrate into pores because of lack of possible pathways. A second migration pathway through the crystalline cocoa butter phase was observed with SAXS. The crystal peak of cocoa butter in form V significantly decreased due to oil migration suggesting that oil migrates through the cocoa butter phase leading to structural changes. This effect was not detected for pure cocoa butter. Oil migration through pores might facilitate the dissolution probably because of higher contact area between lipid migrant and cocoa butter phase. According to the SAXS study, the mechanism for lipid migration in chocolates involves both bulk flow, to be considered as the dominant mechanism in the short term, and chemical migration of lipids through the cocoa butter matrix phase at longer observation times.

Macroscopic observation of migration through cocoa butter in combination with findings from the structural analysis by small angle X-ray scattering and tomography exposed that structure is a key parameter influencing migration in chocolate. Migration rates vary due to change of structure over the storage time.

Figure 7.1 summarizes the relationships between structure and migration of mobile lipids through crystalline fat suspensions based on the experimental results discussed in this thesis. Observation of macroscopic migration in combination with structure analysis by tomography substantiated that structure impacts migration in the investigated systems. However, structure is not a constant but changes due to dissolution and crystallization as shown with visual observations and the SAXS study. Consequently, migration rates are not constant but can be related to a change of structure. Migration of lipids through

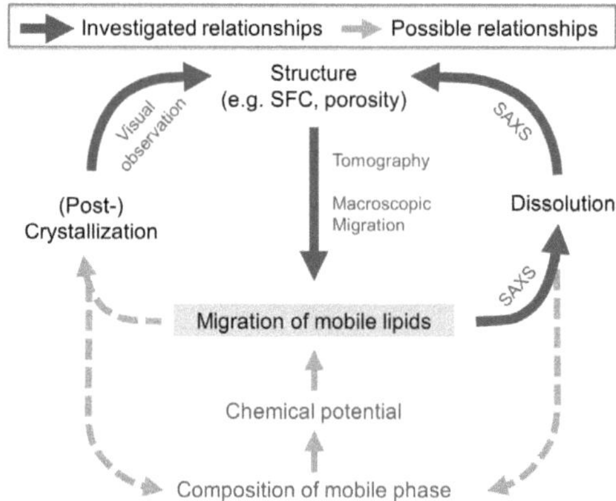

FIGURE 7.1: Overview over structure-migration relationship.

porous glass with an average pore radius of 3.4 nm can be described with Darcy's equation and capillary pressure as driving force. Thereby, the effective cross section for flow is reduced due to sticking of molecules at the inner capillary wall. Further knowledge on the structure of chocolate, such as porosity, tortuosity and capillary size, in the nanometer range are needed, in order to verify the model for migration in chocolate. The effect of storage time and migration on the composition of the mobile phase has not been investigated yet. Following, further research has to analyze the impact of chemical potential on migration. It is likely that crystallization and dissolution both impact the composition of the mobile phase. It is possible that it changes over time because the higher melting fat molecules are prone to the crystal phase forming a denser structure and leaving the lower melting lipids in the mobile phase. Additionally, it is not clearly understood if the mobile lipids migrate as a mixture or if only specific lipids migrate.

Because the material's porous structure plays an important role, this process could be prevented by ensuring a dense and stable structure and minimizing defects in the chocolate matrix, which might act as pathways. Furthermore, the effect of dissolution and chemical migration could be minimized by reducing content of liquid fat. Future research has to show in how far and under which circumstances lipid migration leads to fat bloom formation.

8

Appendix

Example MATLAB code to calculate capillary rise through porous glass with different ratios fh

```matlab
clear all; close all;

% CONSTANTS
g=9.81; % Gravity [m/s^2]
eta = 60*10^-3; % Dynamic viscosity [N*s/m^2]
sigma = 35*10^(-3); % Surface tension of oil [N/m]
rho=925; % Density oil [kg/m^3]
ca = 19.3; % Contact angle [degree]
phi_0 = 0.3; % Volume porosity of the matrix material [-]
phi_i = 0.245; % Volume porosity of the matrix at start of migration [-]
r_0=3.4*10^-9; % Average capillary radius [m]
rl=(3.4-0.25)*10^-9; % Capillary radius for formation of mensicus [m]
tau = 3.6; % tortuosity [-]

% INPUT PARAMETER
% y-axis intercept from regression line of experimental data of h vs. sqrt(t)
u=3.8762;
% Slope from regression line of experimental data of h vs. sqrt(t)
v=4.06029;
fh_slope=power((v/(10^3*sqrt(3600*24)))^2/...
(phi_0/(2*phi_i*eta*r_0^2*tau*rl)*sigma*cosd(ca)),0.25)/r_0;

% Starting time of mgration (due to lag phase)
t0 = power((u/v),2)*3600*24; % [s]
t_max_input=12;
t_max=power(power(t_max_input*(3600*24), 0.5)-sqrt(t0),2);
```

```matlab
n=50; % number of steps to take for time
t_stepsize=(t_max)/(n);
t=[0:t_stepsize:n*t_stepsize]; %[s]


fh_step=10;
z=fh_step+1;
fh_glass=ones(1,10);
fh=[0:1/fh_step:1];


for j=1:z
    for i=1:n+1
    h(j,i)=sqrt(sigma*cosd(ca)*power(r_0*fh(j),4)*phi_0/ ...
    (2*phi_i*eta*r_0^2*tau*rl)*t(i));
    end
end


h_mm=h.*10^3;
sqrt_t_days=sqrt(t./3600/24) + u/v;
t_days = power(sqrt_t_days,2);


% Experimental data from macroscopic migration (oil through porous glass)
t_glass=[t0/3600/24, 5, 8, 12]; % [days]
sqrt_t_glass=power(t_glass,0.5);
h_glass=[0, 5, 8, 10]; % [mm]
```

Capillary pressure

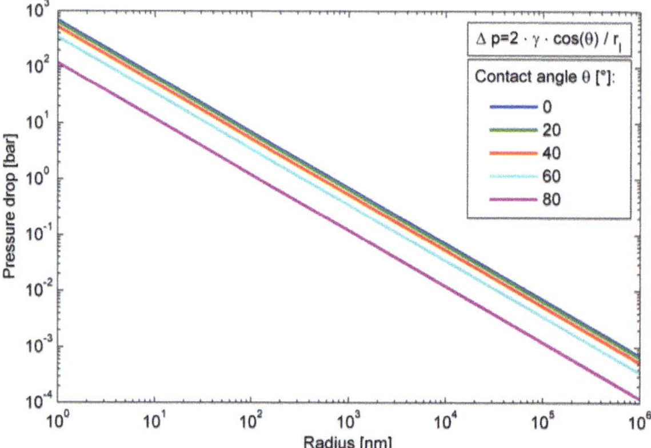

FIGURE 8.1: Pressure drop (Equation 2.8) in capillary with different radius and contact angle between liquid and capillary wall.

Tables with additional simulated and fitting values

TABLE 8.1: Data from simulation of migration height for tempered cocoa butter.

t_0	t_{max}	h_0	r_{ini}	a_{sim}	c_{sim}	a_{ini}	r_{sim}	residual
days	days	mm	nm				nm	mm
0	8	0	500	2.25E-14	2.23E-14	2.039E-08	0.53	5.30E-05
0	8	0	10	2.25E-14	2.23E-14	4.077E-10	0.07	5.30E-05
5	25	0.17	500	1.48E-13	2.28E-14	3.157E-08	1.08	1.88E-04 (*)
5	25	0.17	10	1.601E-13	2.28E-14	6.31E-10	0.16	1.79E-04 (*)
8	20	0.75	500	2.22E-14	2.23E-14	2.039E-08	0.52	5.34E-05
8	20	0.75	10	2.22E-14	2.23E-14	4.077E-10	0.07	5.34E-05
20	50	0.83	500	6.91E-13	2.35E-14	2.039E-08	2.91	3.56E-04
20	50	0.83	10	1.05E-13	2.28E-14	4.077E-10	0.16	1.09E-03
0	50	0	500	6.49E-13	2.35E-14	2.039E-08	2.82	4.83E-04
0	50	0	10	8.65E-14	2.27E-14	4.08E-10	0.15	7.86E-04

TABLE 8.2: Data from simulation of migration height for untempered cocoa butter.

t_0	t_{max}	h_0	r_{ini}	a_{sim}	c_{sim}	a_{ini}	r_{sim}	residual
days	days	mm	nm				nm	mm
0	10	0	500	2.93E-12	2.41E-14	2.04E-08	5.99	3.03E-04
0	10	0	10	4.07E-12	2.39E-14	4.08E-10	1.00	1.81E-04
0	20	0	500	2.52E-12	2.39E-14	2.04E-08	5.56	3.45E-04
0	20	0	10	2.63E-12	3.79E-08	4.08E-10	0.80	3.71E-04
0	25	0	500	2.52E-12	2.39E-14	2.04E-08	5.56	3.45E-04
0	25	0	10	2.63E-12	3.79E-08	4.08E-10	0.80	3.71E-04
20	50	2.58	500	1.23E-12	2.36E-14	2.04E-08	3.89	1.75E-04
20	50	2.58	10	1.10E-12	2.36E-14	4.08E-10	0.52	2.04E-04
0	50	0	500	2.59E-12	2.39E-14	2.04E-08	5.64	3.63E-04
0	50	0	10	2.93E-12	2.38E-14	4.08E-10	0.85	3.11E-04

TABLE 8.3: 4th polynomial fitting to approximate migration heights with time.

Sample	Approximation	constants			
	A0	A1	A2	A3	A4
UUU_down	4.28E-01	4.53E-01	-3.38E-02	1.04E-03	-1.02E-05
UUU_up	1.95E-01	9.69E-02	-1.62E-03	-1.62E-05	9.87E-07
GC1_TU_down	1.15E-01	1.84E-01	3.77E-03	-2.21E-04	2.21E-06
GC1_TU_up	-1.47E-02	1.55E-01	-7.99E-03	2.21E-04	-1.88E-06
GC2_TU_down	4.71E-02	5.12E-01	-3.70E-02	1.33E-03	-1.65E-05
GC2_TU_up	4.50E-02	1.30E-01	-8.73E-03	3.20E-04	-2.84E-06
GC1_UT_down	3.21E-02	-7.15E-02	1.46E-02	-4.74E-04	4.93E-06
GC1_UT_up	3.43E-02	1.01E-01	-9.25E-03	3.42E-04	-3.73E-06
GC2_UT_down	5.60E-03	2.35E-02	1.28E-03	-1.64E-05	2.53E-07
GC2_UT_up	4.90E-02	1.63E-01	-1.02E-02	3.10E-04	-2.91E-06
TTT_down	-1.69E-02	7.80E-02	-3.62E-03	1.46E-04	-1.57E-06
TTT_up	3.01E-02	3.57E-02	-1.28E-03	5.50E-05	-5.72E-07

TABLE 8.4: Residuum for approximated 4th polynomial curve from data. Sample notation as follows: 'System (B, C1 or C2): stained layer (U = untempered; T = tempered) $\rightarrow$ layer into which migration is observed (U = untempered; T = tempered) (direction: up or down)'. Statistical values for migration rates are added: max. = maximum migration rate; min. = minimum migration rate; STD = standard deviation of migration rate.

Sample	Residuum [mm]		Migration rates [mm/day]			
	max.	mean	max.	min.	mean	STD
B: U→U (down)*	0.46	0.16	0.45	0.00	0.10	0.10
B: U→U (up)	0.46	0.20	0.28	0.04	0.09	0.06
C1: T→U (down)*	0.35	0.13	0.21	0.13	0.19	0.02
C1: T→U (up)	0.59	0.25	0.15	0.04	0.07	0.03
C2: T→U (down)*	0.45	0.20	0.51	0.08	0.16	0.11
C2: T→U (up)	0.16	0.09	0.22	0.04	0.10	0.05
C1: U→T (down)	0.29	0.29	0.12	-0.07	0.07	0.04
C1: U→T (up)	0.09	0.05	0.10	0.00	0.04	0.03
C2: U→T (down)	0.08	0.04	0.10	0.02	0.06	0.02
C2: U→T (up)	0.13	0.06	0.16	0.03	0.06	0.03
B: T→T (down)	0.24	0.24	0.09	0.03	0.07	0.02
B: T→T (up)	0.13	0.06	0.05	0.02	0.04	0.01

* The maximum samples height reached before end of observation period. Only the data up to a maximum height of 4 mm considered.

Simulations of hydrodynamic radius for different diffusion coefficients and initial radii for larger radii and diffusion coefficient than in Figure 6.33 and 6.34

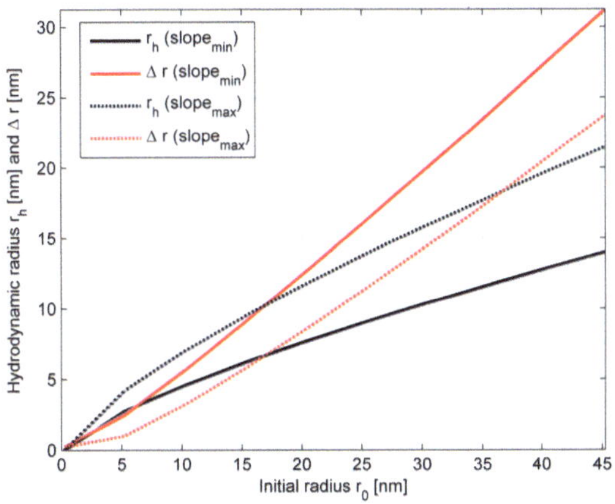

FIGURE 8.2: Change of hydrodynamic radius r_h and difference between initial and hydrodynamic radius $\Delta r = r_0 - r_h$ for different initial radii r_0 for two different effective diffusion coefficients D_{eff} with a slope of 1 mm/day$^{0.5}$ and 2.5 mm/day$^{0.5}$.

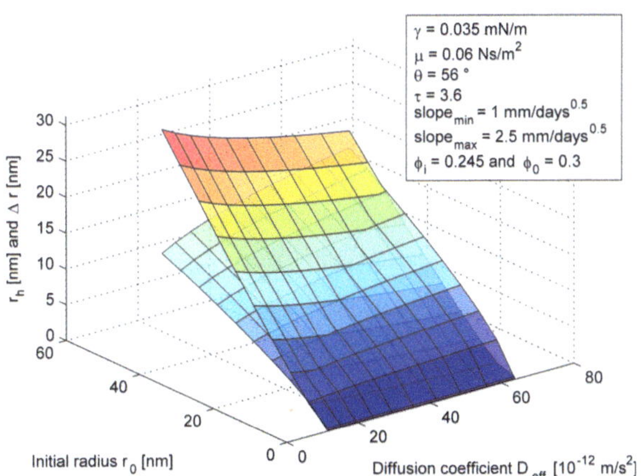

FIGURE 8.3: Change of hydrodynamic radius r_h and difference between initial and hydrodynamic radius $\Delta r = r_0 - r_h$ for different initial radii r_0 with variation of effective diffusion coefficient D_{eff} induced by different slopes in migration height per square root of time.

Bibliography

Acevedo, N. C., Peyronel, F., and Marangoni, A. G. (2011). Nanoscale structure inter-crystalline interactions in fat crystal networks. *Current Opinion in Colloid & Interface Science*, 16(5):374 – 383.

Adam-Berret, M., Boulard, M., Riaublanc, A., and Mariette, F. (2011). Evolution of fat crystal network microstructure followed by NMR. *Journal of Agricultural and Food Chemistry*, 59(5):1767–1773. PMID: 21314096.

Afoakwa, E., Paterson, A., Fowler, M., and Vieira, J. (2008a). Effects of tempering and fat crystallisation behaviour on microstructure, mechanical properties and appearance in dark chocolate systems. *Journal of Food Engineering*, 89(2):128–136.

Afoakwa, E. O., Paterson, A., Fowler, M., and Vieira, J. (2008b). Characterization of melting properties in dark chocolates from varying particle size distribution and composition using differential scanning calorimetry. *Food Research International*, 41(7):751–757.

Afoakwa, E. O., Paterson, A., Fowler, M., and Vieira, J. (2009a). Fat bloom development and structure-appearance relationships during storage of under-tempered dark chocolates. *Journal of Food Engineering*, 91(4):571–581.

Afoakwa, E. O., Paterson, A., Fowler, M., and Vieira, J. (2009b). Influence of tempering and fat crystallization behaviours on microstructural and melting properties in dark chocolate systems. *Food Research International*, 42(1):200 – 209.

Aguilera, J. M. (2005). Why food microstructure? *Journal of Food Engineering*, 67(1–2):3–11.

Aguilera, J. M., Michel, M., and Mayor, G. (2004). Fat migration in chocolate: Diffusion or capillary flow in a particulate solid? - A hypothesis paper. *Journal of Food Science*, 69(7):167 – 174.

Altimiras, P., Pyle, L., and Bouchon, P. (2007). Structure–fat migration relationships during storage of cocoa butter model bars: Bloom development and possible mechanisms. *Journal of Food Engineering*, 80(2):600–610.

Avrami, M. (1939). Kinetics of phase change. I General theory. *The Journal of Chemical Physics*, 7(12):1103–1112.

Avrami, M. (1940). Kinetics of phase change. II Transformation-time relations for random distribution of nuclei. *The Journal of Chemical Physics*, 8(2):212–224.

Avrami, M. (1941). Kinetics of phase change. III granulation, phase change, and microstructure. *The Journal of Chemical Physics*, 9(2):177–184.

Bailey, A. C. and Yates, B. (1970). Anisotropic thermal expansion of pyrolytic graphite at low temperatures. *Journal of Applied Physics*, 41(13):5088–5091.

Beckett, S. T. (2000). Crystallising the fat in chocolate. In Beckett, S. T., editor, *The Science of Chocolate*, pages 85–103. The Royal Society of Chemistry.

Beckett, S. T. (2009). *Traditional Chocolate Making*, pages 1–9. Wiley-Blackwell.

Beckmann, F., Herzen, J., Haibel, A., Müller, B., and Schreyer, A. (2008). High density resolution in synchrotron-radiation-based attenuation-contrast microtomography. *Proceedings SPIE*, 7078:70781D–70781D–13.

Benecke, G., Wagermaier, W., Li, C., Schwartzkopf, M., Flucke, G., Hoerth, R., Zizak, I., Burghammer, M., Metwalli, E., Müller-Buschbaum, P., Trebbin, M., Förster, S., Paris, O., Roth, S. V., and Fratzl, P. (2014). A customizable software for fast reduction and analysis of large X-ray scattering data sets: applications of the new *DPDAK* package to small-angle X-ray scattering and grazing-incidence small-angle X-ray scattering. *Journal of Applied Crystallography*, 47(5):1797–1803.

Bird, R. B., Stewart, W. E., and Lightfoot, E. N. (2007). *Transport Phenomena, revised 2nd edition*. John Wiley & Sons, Inc.

Bouchon, P., Aguilera, J. M., and Pyle, D. L. (2003). Structure oil-absorption relationships during deep-fat frying. *Journal of Food Science*, 68(9):2711–2716.

Brakel, J. V. and Heertjes, P. (1977). Capillary rise in porous media. part I. A problem. *Powder Technology*, 16(1):75 – 81.

Bricknell, J. and Hartel, R. W. (1998). Relation of fat bloom in chocolate to polymorphic transition of cocoa butter. *Journal of the American Oil Chemists' Society*, 75(11):1609–1615.

Bridgman, P. (1923). The thermal conductivity of liquids. *Proceedings of the National Academy of Sciences of the United States of America*, 9:341–345.

Brunello, N., McGauley, S. E., and Marangoni, A. (2003). Mechanical properties of cocoa butter in relation to its crystallization behavior and microstructure. *LWT - Food Science and Technology*, 36(5):525–532.

Bürkle (2011). Viscosity of liquids.

Campos, R., Ollivon, M., and Marangoni, A. G. (2010). Molecular composition dynamics and structure of cocoa butter. *Crystal Growth & Design*, 10(1):205–217.

Cebula, D. J. and Ziegleder, G. (1993). Studies of bloom formation using x-ray diffraction from chocolates after long-term storage. *Lipid / Fett*, 95(9):340–343.

Chapman, G., Akehurst, E., and Wright, W. (1971). Cocoa butter and confectionery fats. Studies using programmed temperature X-ray diffraction and differential scanning calorimetry. *Journal of the American Oil Chemists Society*, 48(12):824–830.

Choi, J. Y., McCarthy, L. K., McCarthy, J. M., and Kim, H. M. (2007). Oil migration in chocolate. *Applied Magnetic Resonance*, 32(1):205–220.

Choi, Y. J., McCarthy, K. L., and McCarthy, M. J. (2005). Oil migration in a chocolate confectionery system evaluated by magnetic resonance imaging. *Journal of Food Science*, 70(5):E312–E317.

Cverna, F. (2002). *ASM ready reference. Thermal properties of metals*. Materials Park, Ohio : ASM International.

Dahlenborg, H., Millqvist-Fureby, A., Bergenståhl, B., and Kalnin, D. (2011). Investigation of chocolate surfaces using profilometry and low vacuum scanning electron microscopy. *Journal of the American Oil Chemists' Society*, 88(6):773–783.

Dahlenborg, H., Millqvist-Fureby, A., Brandner, B. D., and Bergenstahl, B. (2012). Study of the porous structure of white chocolate by confocal raman microscopy. *European Journal of Lipid Science and Technology*, 114(8):919–926.

Darcy, H. (1856). Victor Dalmont, Paris.

Debye, P. and Cleland, R. L. (1959). Flow of liquid hydrocarbons in porous vycor. *Journal of Applied Physics*, 30(6):843–849.

Dettmeyer, R. B. (2011). Staining techniques and microscopy. In Dettmeyer, R. B., editor, *Forensic Histopathology: Fundamentals and Perspectives*, pages 17–35. Springer Berlin Heidelberg.

DIN ISO 7991 (1998). Bestimmung des mittleren thermischen Längenausdehnungskoeffizienten.

Duck, W. (1964). The measurement of unstable fat in finished chocolate. *The Manufacturing Confectioner*, June 1964:67 – 72.

El-Mallah, M. H. and Megahed, M. G. (1998). Studies on cocoa butter-replacer mixtures suitable for the local chocolate production. *Grasas y Aceites*, 49(5-6):446–449.

Engmann, J. and Mackley, M. (2006). Semi-solid processing of chocolate and cocoa butter: Modelling rheology and microstructure changes during extrusion. *Food and Bioproducts Processing*, 84(2):102–108.

Frazier, A. and Hartel, R. (2012). Bloom on chocolate chips baked in cookies. *Food Research International*, 48(2):380 – 386.

Fries, N. and Dreyer, M. (2008). An analytic solution of capillary rise restrained by gravity. *Journal of Colloid and Interface Science*, 320(1):259 – 263.

G. Ziegleder, J. Geier-Greguska, J. G. (1994). HPLC-Analyse von Fettreif. *Fat Science Technology*, 96(10):390–394.

Galdámez, J. R., Szlachetka, K., Duda, J. L., and Ziegler, G. R. (2009). Oil migration in chocolate: A case of non-fickian diffusion. *Journal of Food Engineering*, 92(3):261 – 268.

Gebhardt, R., Burghammer, M., Riekel, C., Kulozik, U., and Müller-Buschbaumm, P. (2010a). Investigation of surface modification of casein films by rennin enzyme action using micro-beam grazing incidence small angle X-ray scattering. *Dairy Science and Technology*, 90(1):75–86.

Gebhardt, R., Roth, S. V., Burghammer, M., Riekel, C., Tolkach, A., Kulozik, U., and Müller-Buschbaum, P. (2010b). Structural changes of casein micelles in a rennin gradient film with simultaneous consideration of the film morphology. *International Dairy Journal*, 20(3):203 – 211.

Gebhardt, R., Vendrely, C., and Kulozik, U. (2011). Structural characterization of casein micelles, shape changes during film formation. *Journal of Physics Condensed Matter*, 23(44):444201.

Ghosh, V., Ziegler, G. R., and Anantheswaran, R. C. (2002). Fat, moisture, and ethanol migration through chocolates and confectionary coatings. *Critical Reviews in Food Science and Nutrition*, 42(6):583–626.

Green, N. L. and Rousseau, D. (2014). Oil migration: Novel approaches to a familiar problem. *Lipid Technology*, 26(11-12):243–245.

Greving, I., Wilde, F., Ogurreck, M., Herzen, J., Hammel, J. U., Hipp, A., Friedrich, F., Lottermoser, L., Dose, T., Burmester, H., Müller, M., and Beckmann, F. (2014). P05 imaging beamline at PETRA III: first results. *Proceedings SPIE*, 9212:92120O-1-92120O-8.

Grüner, S., Hermes, H. E., Schillinger, B., Egelhaaf, S. U., and Huber, P. (2016). Capillary rise dynamics of liquid hydrocarbons in mesoporous silica as explored by gravimetry, optical and neutron imaging: Nano-rheology and determination of pore size distributions from the shape of imbibition fronts. *Colloids and Surfaces A: Physicochemical and Engineering Aspects*, 496:13 – 27. Characterization of porous materials: from Angstroms to millimeters – {VII}.

Grüner, S., Hofmann, T., Wallacher, D., Kityk, A. V., and Huber, P. (2009). Capillary rise of water in hydrophilic nanopores. *Phys. Rev. E*, 79:067301.

Guiheneuf, T. M., Couzens, P. J., Wille, H.-J., and Hall, L. D. (1997). Visualisation of liquid triacylglycerol migration in chocolate by magnetic resonance imaging. *Journal of the Science of Food and Agriculture*, 73(3):265–273.

Haibel, A., Beckmann, F., Dose, T., Herzen, J., Ogurreck, M., Müller, M., and Schreyer, A. (2010). Latest developments in microtomography and nanotomography at PETRA III. *Powder Diffraction*, 25:161–164.

Haighton, A. J. (1965). Worksoftening of margarine and shortening. *Journal of the American Oil Chemists' Society*, 42(1):27–30.

Hartel, R. W. (1999). Chocolate: Fat bloom during storage. *The Manufacturing Confectioner*, 79(5):89–99.

Hernqvist, L. (1990). Polymorphism of triglycerides - A crystallographic review. *Food Structure*, 9(1):39–44.

Himawan, C., Starov, V., and Stapley, A. (2006). Thermodynamic and kinetic aspects of fat crystallization. *Advances in Colloid and Interface Science*, 122(1–3):3 – 33. ECIC-XVII, {XVIIth} European Chemistry at Interfaces Conference, 27 June-1 July, 2005, Department of Chemical Engineering, Loughborough University.

Hodge, S. M. and Rousseau, D. (2002). Fat bloom formation and characterization in milk chocolate observed by atomic force microscopy. *Journal of the American Oil Chemists' Society*, 79(11):1115–1121.

Holt, C., de Kruif, C., Tuinier, R., and Timmins, P. (2003). Substructure of bovine casein micelles by small-angle X-ray and neutron scattering. *Colloids and Surfaces A: Physicochemical and Engineering Aspects*, 213(2–3):275 – 284.

Hondoh, H., Yamasaki, K., Ikutake, M., and Ueno, S. (2016). Visualization of oil migration in chocolate using scanning electron microscopy–energy dispersive X-ray spectroscopy. *Food Structure*, 8:8 – 15.

Huyghebaert, A. and Hendrickx, H. (1971). Polymorphism of cocoa butter, shonw by differential scanning calorimetry. *Lebensmittel - Wissenschaft und Technologie*, 4:59–63.

James, B. J. and Smith, B. G. (2009). Surface structure and composition of fresh and bloomed chocolate analysed using X-ray photoelectron spectroscopy, cryo-scanning electron microscopy and environmental scanning electron microscopy. *LWT - Food Science and Technology*, 42(5):929–937.

Kadivar, S., Clercq, N. D., Mokbul, M., and Dewettinck, K. (2016). Influence of enzymatically produced sunflower oil based cocoa butter equivalents on the phase behavior of cocoa butter and quality of dark chocolate. *{LWT} - Food Science and Technology*, 66:48 – 55.

Khan, R. S. and Rousseau, D. (2006). Hazelnut oil migration in dark chocolate – kinetic, thermodynamic and structural considerations. *European Journal of Lipid Science and Technology*, 108(5):434–443.

Kinta, Y. and Hatta, T. (2005). Composition and structure of fat bloom in untempered chocolate. *Journal of Food Science*, 70(7):s450–s452.

Kinta, Y. and Hatta, T. (2006). Composition, structure, and color of fat bloom due to the partial liquefaction of fat in dark chocolate. *Journal of the American Oil Chemists' Society*, 84(2):107–115.

Le Reverend, B. J. D., Fryer, P. J., and Bakalis, S. (2009). Modelling crystallization and melting kinetics of cocoa butter in chocolate and application to confectionery manufacturing. *Soft Matter*, 5:891–902.

Lee, W. L., McCarthy, M. J., and McCarthy, K. L. (2010). Oil migration in 2-component confectionery systems. *Journal of Food Science*, 75(1):E83–E89.

Liang, B., Sebright, J. L., Shi, Y., Hartel, R. W., and Perepezko, J. H. (2006). Approaches to quantification of microstructure for model lipid systems. *Journal of the American Oil Chemists' Society*, 83(5):389–399.

Liang, B., Shi, Y., and Hartel, R. W. (2008). Correlation of rheological and microstructural properties in a model lipid system. *Journal of the American Oil Chemists' Society*, 85(5):397–404.

Lin, M. Y., Abeles, B., Huang, J. S., Stasiewski, H. E., and Zhang, Q. (1992). Viscous flow and diffusion of liquids in microporous glasses. *Phys. Rev. B*, 46:10701–10705.

Lipp, M. and Anklam, E. (1998). Review of cocoa butter and alternative fats for use in chocolate - Part A. compositional data. *Food Chemistry*, 62(1):73 – 97.

Lohman, M. H. and Hartel, R. W. (1994). Effect of milk fat fractions on fat bloom in dark chocolate. *Journal of the American Oil Chemists' Society*, 71(3):267–276.

Loisel, C., Keller, G., Lecq, G., Bourgaux, C., and Ollivon, M. (1998). Phase transitions and polymorphism of cocoa butter. *Journal of the American Oil Chemists' Society*, 75(4):425–439.

Loisel, C., Lecq, G., Ponchel, G., Keller, G., and Ollivon, M. (1997). Fat bloom and chocolate structure studied by mercury porosimetry. *Journal of Food Science*, 62(4):781–788.

Lonchampt, P. and Hartel, R. W. (2004). Fat bloom in chocolate and compound coatings. *European Journal of Lipid Science and Technology*, 106(4):241–274.

Lovegren, N., Gray, M., and Feuge, R. (1976). Effect of liquid fat on melting point and polymorphic behavior of cocoa butter and a cocoa butter fraction. *Journal of the American Oil Chemists Society*, 53(3):108–112.

Lucas, R. (1918). Über das Zeitgesetz des kapillaren Aufstiegs von Flüssigkeiten. *Kolloid-Zeitschrift*, 23(1):15–22.

Lügger, S. K., Wilde, F., Dülger, N., Reinke, L. M., Kozhar, S., Beckmann, F., Greving, I., Vieira, J., Heinrich, S., and Palzer, S. (2016). Synchrotron x-ray microtomography of the interior microstructure of chocolate. *Proceedings SPIE, Developments in X-Ray Tomography X*, 9967:99670N.

Lutton, E. S. (1950). Review of the polymorphism of saturated even glycerides. *Journal of the American Oil Chemists Society*, 27(7):276–281.

Maleky, F. and Marangoni, A. (2011a). Nanoscale effects on oil migration through triacylglycerol polycrystalline colloidal networks. *Soft Matter*, 7:6012–6024.

Maleky, F. and Marangoni, A. (2011b). Thermal and mechanical properties of cocoa butter crystallized under an external laminar shear field. *Crystal Growth & Design*, 11(6):2429–2437.

Maleky, F., McCarthy, K. L., McCarthy, M. J., and Marangoni, A. G. (2012). Effect of cocoa butter structure on oil migration. *Journal of Food Science*, 77(3):E74–E79.

Marangoni, A. G. (2000). Elasticity of high-volume-fraction fractal aggregate networks: A thermodynamic approach. *Phys. Rev. B*, 62:13951–13955.

Marangoni, A. G. and McGauley, S. E. (2003). Relationship between crystallization behavior and structure in cocoa butter. *Crystal Growth & Design*, 3(1):95–108.

Marangoni, A. G. and Rogers, M. A. (2003). Structural basis for the yield stress in plastic disperse systems. *Applied Physics Letters*, 82(19):3239–3241.

Marty, S., Baker, K., Dibildox-Alvarado, E., Rodrigues, J. N., and Marangoni, A. G. (2005). Monitoring and quantifying of oil migration in cocoa butter using a flatbed scanner and fluorescence light microscopy. *Food Research International*, 38(10):1189 – 1197.

Marty, S., Baker, K. W., and Marangoni, A. G. (2009). Optimization of a scanner imaging technique to accurately study oil migration kinetics. *Food Research International*, 42(3):368 – 373.

Marty, S. and Marangoni, A. G. (2009). Effects of cocoa butter origin, tempering procedure, and structure on oil migration kinetics. *Crystal Growth & Design*, 9(10):4415–4423.

McCarthy, M. J., Reid, D. S., and Wei, D. (2003). Fat bloom in chocolate. *The Manufacturing Confectioner*, September 2003:89 – 93.

Mehrle, Y. E. (2007). *Solidification and Contraction of Confectionery Systems in Rapid Cooling Processing*. PhD thesis, ETH Zurich.

Merken, G. V. and Vaeck, S. V. (1980). Etude du polymorphisme du beurre de cacao par calorimetrie DSC. *Lebensmittel - Wissenschaft und Technologie*, 13:314–317.

Metin, S. and Hartel, R. W. (2005). *Crystallization of Fats and Oils*, chapter 2. John Wiley and Sons, Inc.

Miquel, M. E., Carli, S., Couzens, P. J., Wille, H.-J., and Hall, L. D. (2001). Kinetics of the migration of lipids in composite chocolate measured by magnetic resonance imaging. *Food Research International*, 34(9):773 – 781.

Molenda, M., Stasiak, M., Moya, M., Ramirez, A., Horabik, J., and Ayuga, F. (2006). Testing mechanical properties of food powders in two laboratories - degree of consistency of results. *International Agrophysics*, 20(1):37–45.

Narine, S. S., Humphrey, K. L., and Bouzidi, L. (2006). Modification of the Avrami model for application to the kinetics of the melt crystallization of lipids. *Journal of the American Oil Chemists' Society*, 83(11):913–921.

Narine, S. S. and Marangoni, A. G. (1999a). The difference between cocoa butter and salatrim lies in the microstructure of the fat crystal network. *Journal of the American Oil Chemists' Society*, 76(1):7–13.

Narine, S. S. and Marangoni, A. G. (1999b). Fractal nature of fat crystal networks. *Phys. Rev. E*, 59:1908–1920.

Narine, S. S. and Marangoni, A. G. (1999c). Mechanical and structural model of fractal networks of fat crystals at low deformations. *Physical Review E*, 60:6991–7000.

Otto, J. and Thomas, W. (1963). Die thermische Ausdehnung von Quarzglas im Temperaturbereich von 0 bis 1060 C. *Zeitschrift für Physik*, 175(3):337–344.

Palzer, S. (2005). The effect of glass transition on the desired and undesired agglomeration of amorphous food powders. *Chemical Engineering Science*, 60(14):3959 – 3968.

Patsioura, A., Vauvre, J.-M., Kesteloot, R., Smith, P., Trystram, G., and Vitrac, O. (2016). Mechanisms of oil uptake in french fries. In Singh, J. and Kaur, L., editors, *Advances in Potato Chemistry and Technology (Second Edition)*, chapter 17, pages 503 – 526. Academic Press, San Diego, second edition edition.

Pore, M., Seah, H. H., Glover, J. W. H., Holmes, D. J., Johns, M. L., Wilson, D. I., and Moggridge, G. D. (2009). In-situ X-ray studies of cocoa butter droplets undergoing simulated spray freezing. *Journal of the American Oil Chemists' Society*, 86(3):215–225.

Porod, G. (1982). General theory. In Glatter, O. and Kratky, O., editors, *Small Angle X-Ray Scattering*, chapter 2, pages 17–51. Academic Press: New York.

Quevedo, R., Brown, C., Bouchon, P., and Aguilera, J. (2005). Surface roughness during storage of chocolate: Fractal analysis and possible mechanisms. *Journal of the American Oil Chemists' Society*, 82(6):457–462.

Qureshi, J. A., Buschman, L. L., Throne, J. E., and Ramaswamy, S. B. (2004). Oil-soluble dyes incorporated in meridic diet of diatraea grandiosella (lepidoptera: Crambidae) as markers for adult dispersal studies. *Journal of Economic Entomology*, 97(3):836–845.

Ramos, K. and Bahr, D. (2007). Mechanical behavior assessment of sucrose using nanoindentation. *Journal of Materials Research*, 22:2037–2045.

Ramírez, A., Moya, M., and Ayuga, F. (2009). Determination of the mechanical properties of powdered agricultural products and sugar. *Particle and Particle Systems Characterization*, 26(4):220–230.

Reid, A. C., Lua, R. C., Garcia, R. E., Coffman, V. R., and Langer, S. A. (2009). Modelling microstructures with OOF2. *International Journal of Materials and Product Technology*, 35(3/4):361–373.

Reinke, S. K., Hauf, K., Vieira, J., Heinrich, S., and Palzer, S. (2015a). Changes in contact angle providing evidence for surface alteration in multi-component solid foods. *Journal of Physics D: Applied Physics*, 48(46):464001.

Reinke, S. K., Roth, S. V., Santoro, G., Vieira, J., Heinrich, S., and Palzer, S. (2015b). Tracking structural changes in lipid-based multicomponent food materials due to oil migration by microfocus small-angle X-ray scattering. *ACS Applied Materials & Interfaces*, 7(18):9929–9936. PMID: 25894460.

Reinke, S. K., Wilde, F., Kozhar, S., Beckmann, F., Vieira, J., Heinrich, S., and Palzer, S. (2015c). Synchrotron X-ray microtomography reveals interior microstructure of multicomponent food materials such as chocolate. *Journal of Food Engineering*, pages –.

Ribeiro, A. P. B., Basso, R. C., and Kieckbusch, T. G. (2013). Effect of the addition of hardfats on the physical properties of cocoa butter. *European Journal of Lipid Science and Technology*, 115(3):301–312.

Rønholt, S., Mortensen, K., and Knudsen, J. C. (2013). The effective factors on the structure of butter and other milk fat-based products. *Comprehensive Reviews in Food Science and Food Safety*, 12(5):468–482.

Rothkopf, I. and Danzl, W. (2015). Changes in chocolate crystallization are influenced by type and amount of introduced filling lipids. *European Journal of Lipid Science and Technology*, 117(11):1714–1721.

Rousseau, D. (2006). On the porous mesostructure of milk chocolate viewed with atomic force microscopy. *LWT - Food Science and Technology*, 39(8):852–860.

Rousseau, D. and Smith, P. (2008). Microstructure of fat bloom development in plain and filled chocolate confections. *Soft Matter*, 4:1706–1712.

Rousseau, D. and Sonwai, S. (2008). Influence of the dispersed particulate in chocolate on cocoa butter microstructure and fat crystal growth during storage. *Food Biophysics*, 3(2):273–278.

Sato, K. (1993). Polymorphic transformations in crystal growth. *J. Phys. D Awl. Phys.*, 26:B77–B84.

Sato, K. (1999). Solidification and phase transformation behaviour of food fats — a review. *Lipid / Fett*, 101(12):467–474.

Sato, K. (2001). Crystallization behaviour of fats and lipids - A review. _Chemical Engineering Science_, 56(7):2255 – 2265. Industrial Crystallisation.

Sato, K., Arishima, T., Wang, Z. H., Ojima, K., Sagi, N., and Mori, H. (1989). Polymorphism of pop and sos. i. occurrence and polymorphic transformation. _Journal of the American Oil Chemists' Society_, 66(5):664–674.

Schindelin, J., Arganda-Carreras, I., Kaynig, E. F. V., Longair, M., Pietzsch, T., Preibisch, S., Rueden, C., Saalfeld, S., Schmid, B., Tinevez, J.-Y., White, D. J., Hartenstein, V., Eliceiri, K., Tomancak, P., and Cardona, A. (2012). Fiji: an open-source platform for biological-image analysis. _Nature methods_, 9(7):676–682.

Schmid, B., Schindelin, J., Cardona, A., Longair, M., and Heisenberg, M. (2010). A high-level 3D visualization API for Java and ImageJ. _BMC Bioinformatics_, 11(1):1–7.

Shukla, A., Narayanan, T., and Zanchi, D. (2009). Structure of casein micelles and their complexation with tannins. _Soft Matter_, 5:2884–2888.

Smith, K. W., Bhaggan, K., Talbot, G., and Malssen, K. F. (2011). Crystallization of fats: Influence of minor components and additives. _Journal of the American Oil Chemists' Society_, 88(8):1085–1101.

Smith, K. W., Cain, F. W., and Talbot, G. (2007). Effect of nut oil migration on polymorphic transformation in a model system. _Food Chemistry_, 102(3):656 – 663.

Smith, P. R. and Dahlman, A. (2005). The use of atomic force microscopy to measure the formation and development of chocolate bloom in pralines. _Journal of the American Oil Chemists' Society_, 82(3):165–168.

Stalder, A., Melchior, T., Müller, M., Sage, D., Blu, T., and Unser, M. (2010). Low-bond axisymmetric drop shape analysis for surface tension and contact angle measurements of sessile drops. _Colloids and Surfaces A: Physicochemical and Engineering Aspects_, 364(1-3):72–81.

Stukan, M. R., Ligneul, P., Crawshaw, J. P., and Boek, E. S. (2010). Spontaneous imbibition in nanopores of different roughness and wettability. _Langmuir_, 26(16):13342–13352.

Svanberg, L., Ahrné, L., Lorén, N., and Windhab, E. (2011a). Effect of sugar, cocoa particles and lecithin on cocoa butter crystallisation in seeded and non-seeded chocolate model systems. _Journal of Food Engineering_, 104(1):70–80.

Svanberg, L., Ahrné, L., Lorén, N., and Windhab, E. (2011b). Effect of pre-crystallization process and solid particle addition on cocoa butter crystallization and resulting microstructure in chocolate model systems. *Procedia Food Science*, 1:1910 – 1917. 11th International Congress on Engineering and Food (ICEF11).

Svanberg, L., Ahrné, L., Lorén, N., and Windhab, E. (2011c). Effect of pre-crystallization process and solid particle addition on microstructure in chocolate model systems. *Food Research International*, 44(5):1339 – 1350.

Svanberg, L., Lorén, N., Ahrné, L., and Windhab, E. (2013). Pre-crystallization to control the fat crystal structure and its relation to storage stability in dark chocolate. *InsideFood Symposium*, 9-12 April 2013. InsideFood Symposium Leuven, Belgium.

Svanberg, L., Lorén, N., and Ahrné, L. (2012). Chocolate swelling during storage caused by fat or moisture migration. *Journal of Food Science*, 77(11):E328–E334.

Tabouret, T. (1987). Technical note: Detection of fat migration in a confectionery product. *International Journal of Food Science & Technology*, 22(2):163–167.

Talbot, G. (2009a). Chocolate temper. In Beckett, S. T., editor, *Industrial Chocolate Manufacture and Use*, pages 261–275. Wiley-Blackwell.

Talbot, G. (2009b). Vegetable fats. In Beckett, S. T., editor, *Industrial Chocolate Manufacture and Use*, pages 415–433. Wiley-Blackwell.

Talbot, G., Smith, K. W., and Cain, F. W. (2006). Oil migration - minimization of extent and effects. *The Manufacturing Confectioner*, March 2006:63–66.

Tempel, M. (1958). Rheology of plastic fats. *Rheologica Acta*, 1(2):115–118.

Thurner, P., Beckmann, F., and Müller, B. (2004). An optimization procedure for spatial and density resolution in hard x-ray micro-computed tomography. *Nuclear Instruments and Methods in Physics Research Section B: Beam Interactions with Materials and Atoms*, 225(4):599 – 603.

Timms, R. E. (2002). Oil and fat interactions, theory, problems and solutions. *The Manufacturing Confectioner*, June 2002:50–64.

Tisoncik, M. (2013). Chocolate fat bloom. *The Manufacturing Confectioner*, April 2013:65 – 68.

Torbica, A., Jovanovic, O., and Pajin, B. (2006). The advantages of solid fat content determination in cocoa butter and cocoa butter equivalents by the Karlshamns method. *European Food Research and Technology*, 222(3-4):385–391.

Trebbin, M., Steinhauser, D., Perlich, J., Buffet, A., Roth, S. V., Zimmermann, W., Thiele, J., and Förster, S. (2013). Anisotropic particles align perpendicular to the flow direction in narrow microchannels. *Proceedings of the National Academy of Sciences*, 110(17):6706–6711.

Van der Weeën, P., Clercq, N. D., Baetens, J. M., Delbaere, C., Dewettinck, K., and Baets, B. D. (2013). A discrete stochastic model for oil migration in chocolate-coated confectionery. *Journal of Food Engineering*, 119(3):602 – 610.

van Malssen, K., Peschar, R., and Schenk, H. (1996). Real-time X-ray powder diffraction investigations on cocoa butter. ii. the relationship between melting behavior and composition of beta-cocoa butter. *Journal of the American Oil Chemists' Society*, 73(10):1217–1223.

van Malssen, K., van Langevelde, A., Peschar, R., and Schenk, H. (1999). Phase behavior and extended phase scheme of static cocoa butter investigated with real-time X-ray powder diffraction. *Journal of the American Oil Chemists' Society*, 76(6):669–676.

van Mechelen, J. B., Peschar, R., and Schenk, H. (2006a). Structures of mono-unsaturated triacylglycerols. I. The β_1 polymorph. *Acta Crystallographica Section B*, 62(6).

van Mechelen, J. B., Peschar, R., and Schenk, H. (2006b). Structures of mono-unsaturated triacylglycerols. II. The β_2 polymorph. *Acta Crystallographica Section B*, 62(6):1131–1138.

Walter, P. and Cornillon, P. (2002). Lipid migration in two-phase chocolate systems investigated by NMR and DSC. *Food Research International*, 35(8):761 – 767.

Wang, F., Liu, Y., Shan, L., Jin, Q., Wang, X., and Li, L. (2010). Blooming in cocoa butter substitutes based compound chocolate: Investigations on composition, morphology and melting behavior. *Journal of the American Oil Chemists' Society*, 87(10):1137–1143.

Washburn, E. W. (1921). The dynamics of capillary flow. *Physical Review*, 17:273–283.

Wille, R. and Lutton, E. (1966). Polymorphism of cocoa butter. *Journal of the American Oil Chemists Society*, 43(8):491–496.

Wohlmuth, E. G. (2009). Recipes. In Beckett, S. T., editor, *Industrial Chocolate Manufacture and Use*, pages 434–450. Wiley-Blackwell.

Young, T. (1805). An essay on the cohesion of fluids. *Philosophical Transactions of the Royal Society of London*, 95:65–87.

Zeng, Y., Braun, P., and Windhab, E. (2002). Tempering - continuous precrystallization of chocolate with seed cocoa butter crystal suspension. *The Manufacturing Confectioner*, 4:71–80.

Zeng, Y. and Windhab, E. J. (1999). Kontinuierliche Vorkristallisation von Schokoladen durch Animpfen mit hochstabilen βVI-Kakaobutterkristallen. *Chemie Ingenieur Technik*, 71(9):1008–1009.

Ziarani, A. S. and Aguilera, R. (2012). Knudsen's permeability correction for tight porous media. *Transport in Porous Media*, 91(1):239–260.

Ziegleder, G. (1990). DSC-Thermoanalyse und Kinetik der Kristallisation von Kakaobutter. *Lipid / Fett*, 92(12):481–485.

Ziegleder, G., Moser, C., and Geier-Greguska, J. (1996a). Kinetik der Fettmigration in Schokoladenprodukten Teil I: Grundlagen und Analytik. *Lipid / Fett*, 98(6):196–199.

Ziegleder, G., Moser, C., and Geier-Greguska, J. (1996b). Kinetik der Fettmigration Teil II: Einfluß von Lagertemperatur, Diffusionskoeffizient, Festfettgehalt. *Lipid / Fett*, 98(7-8):253–256.

Ziegleder, G. and Schwingshandl, I. (1998). Kinetics of fat migration within chocolate products. Part III: fat bloom. *Lipid / Fett*, 100(9):411–415.

Ziegler, G. (2007). Oil migration in chocolate. *The Manufacturing Confectioner*, November 2007:51–55.